A Number for your Thoughts

Facts and Speculations about Numbers
from Euclid to the latest Computers

A NUMBER FOR
YOUR THOUGHTS

Malcolm E Lines

© Adam Hilger Ltd 1986

British Library Cataloguing in Publication Data

Lines, Malcolm E.
   A number of your thoughts: facts and
   speculations about numbers from Euclid to the
   latest computers.
   1. Numbers, Theory of
   I. Title
   512'.7   QA241

   ISBN 0-85274-495-1

First published 1986
Reprinted 1988, 1993

Published by Institute of Physics Publishing, wholly owned by the Institute of Physics, London

Institute of Physics Publishing, Techno House, Redcliffe Way, Bristol BS1 6NX, England

US Editorial Office: Institute of Physics Publishing, The Public Ledger Building, Suite 1035, Independence Square, Philadelphia, PA 19106

Printed in Great Britain by J W Arrowsmith Ltd, Bristol

TABLE OF CONTENTS

1. COUNTING

From prehistoric times man has had the need to count. The stone-age hunter or hunting scout would doubtless have found it of great use to be able to give his hunting colleagues some indication of the number of animals he had located, in addition to their kind and approximate location. Although terms such as one, few, and many, may well have sufficed for a while, a more precise counting scheme would be needed eventually, perhaps for bartering, and some concept of number does seem to be possessed by even the most primitive tribes today. Counting, of course, can be performed without the *verbal* possession of number words. This can be achieved, for example, by placing the objects to be counted in a one-to-one correspondence with fingers, toes, or 'counting stones', but words for the most commonly occurring numbers (usually the smallest) are obviously convenient, and seem to have developed at quite an early stage in all forms of human society.

In order to proceed to large numbers in the counting process it soon becomes clear that some grouping arrangement is highly desirable. Thus, the number twenty three is much more conveniently recorded by two 'marking-stones' designating tens, and three perhaps smaller ones designating units, than by twenty three separate 'unit' stones. The grouping number is, in modern usage, referred to as the *base* of the counting system. It need not be equal to ten, of course, and systems based on five, twenty, and even sixty, have occurred in other cultures. Indeed, remnants of such systems are still with us today in the measurement of time (hours, minutes, and seconds - with base sixty) and in the words dozen (base twelve) and score (base twenty). It is even possible to evolve multi-base counting systems - the Mayans used one - and such systems abound among the English units of measurement which answer the question of 'how much?' rather than 'how many?'. Some readers may still recall the tribulations of working out the old English money system of twelve pence in one shilling and twenty shillings in one pound before it was mercifully decimalized in the early 1970's. Others may be more familiar with at least most of the units of weight such as tons, hundred-weights, stones, pounds, and ounces, but possibly without having a precise recollection of how many of these happen to be in one of those.

Doing arithmetic in these multi-base counting systems can be quite tricky, although familiarity helps to a surprising degree, and when the English currency was finally decimalized, many of the older generation found the new simplified system quite confusing and continued to convert everything back to the old multi-base pounds, shillings, and pence before deciding on the advisability of a particular purchase. Fundamentally, however, the single- base system is the simplest, and such a system with base ten is used almost universally for counting today. For this reason the present book can very largely be restricted to this system alone. It is referred to as the *decimal* system of counting, and the choice of the number ten presumably arose from counting on the fingers, with the word 'digit' for any numeral between 0 and 9 seemingly attesting to this fact.

Numbers are an abstract concept and have no physical form. I cannot therefore write down the number 5. But you just did, I hear you say. Well, not really - I wrote down a particular mark (called a numeral) to represent it. Had I been a Roman I should have written V. There is fortunately nothing absolute about any one representation, and the fundamental properties of numbers are not at all dependent on the notation used. The fact that 5 is not exactly divisible by 2 is still true if I think of it as V divided by II. It is also just as true if I use a different base or grouping terminology. This means that if these properties can be demonstrated in our own familiar base-ten system, then we do not have to worry further about verifying them in other counting systems.

Counting in groups of ten, with the symbols 0,1,2,3,4,5,6,7,8,9, is so ingrained in us that the fact that it is quite an arbitrary choice comes almost as a surprise. That one can count quite happily in systems with other bases, right down to 'base-two', which uses only the two symbols 0 and 1, is something of an alien concept to most of us. After all, counting in groups of ten (or to the base ten as we should more formally say) has been a common procedure in many civilizations since the early Egyptians at least. On the other hand, the precise system which we use today, and so take for granted, is of much more recent origin. It contains within it one of the most important inventions ever made, a property which all earlier counting systems, even those using the base ten, did not have. Once again familiarity breeds contempt, and you may well be wondering what attribute of our simple counting system, which somehow seems so natural, could possibly deserve such an accolade. Wouldn't anyone, you may feel, who chose to count in tens, proceed roughly as we do with possibly different symbols for the numerals? The answer is almost certainly not.

The most natural way to count in groups of ten is first to choose symbols for the first nine digits, and then to choose other symbols to represent ten, twenty, etc., up to ninety, and still others for one hundred, two hundred, and so on. Roman numerals, with which we are all acquainted to some degree, are just such an example. They are based in groups of ten as follows:

Units: I, II, III, IV, V, VI, VII, VIII, IX
Tens: X, XX, XXX, XL, L, LX, LXX, LXXX, XC
Hundreds: C, CC, CCC, CD, D, DC, DCC, DCCC, CM
Thousands: M, MM, MMM, MMMM, MMMMM, and so on.

Thus the number 8888 becomes MMMMMMMMDCCCLXXXVIII in Roman numerals, and a routine shopping list for our everyday Roman Centurian might look something like this:

III pairs sandals	III x VII	equals	XXI	den.
IV tunics	IV x IX	equals	XXXVI	den.
I ceremonial toga	I x XL	equals	XL	den.
II plumed helmets	II x XVIII	equals	XXXVI	den.
I sword (regular)	I x XXVIII	equals	XXVIII	den.
I shield	I x XXXIX	equals	XXXIX	den.
	TOTAL	equals	CC	den.

where den. stands for denarius, which was a Roman unit of currency.

Now I do not known whether the cost of living during any period of the Roman empire was such as to make these values realistic, but it is quite possible since rampant inflation was as much a part of Roman lives as it is of our own. The point which we are trying to make does not depend on this of course; it is that checking this shopping list does seem to be a bit tricky without converting it to more familiar numbers. Similarly, multiplying that Roman equivalent of 8888 set out above by say IX or XI seems to be even more difficult. Are these difficulties due only to our lack of familiarity with the system, or is it more than that? Well, our lack of familiarity is certainly no help, but there is indeed a fundamental difficulty with the Roman system over and above this. We should probably sense it first in this way; there are no units columns and no tens columns. Now it is true that methods can be devised for putting the Roman addition and multiplication into some kind of column format (although there is no evidence that the Romans actually did this) which, when combined with special rules for transfering from some columns to others, enable these tasks to be carried out for relatively small numbers. For large numbers, however, the situation is so bad that to represent a million, let alone multiply it by anything, a Roman would have to fill several pages of this book entirely with M's. It is true that more and more letters could be introduced to represent larger and larger groups of ten, but then the system itself rapidly gets out of hand in any case.

The arithmetic associated with these kinds of problems is eased a little bit if we have just one symbol for each numeral between 1 and 9, and then use a separate set of symbols to designate whether we are dealing with tens, hundreds, thousands, etc. For example, if we designate thousands by M, hundreds by C, and tens by X, to follow the Roman precedent, then the number 8888 referred to above would now appear as 8M8C8X8, or its equivalent with the numeral 8 replaced by whatever symbol might have been chosen to represent eight units. In fact, just such grouping systems as these were used at one time by the Chinese and the Japanese (with suitably oriental characters for the numerals and the group symbols of course). But with this system there are still significant problems remaining concerning the writing of large numbers. For each larger group of ten a new symbol is required. Millions are not too bad (requiring five group symbols) but astronomical calculations would certainly tax this system.

One important step remains to take us over to our familiar modern system. It is the all-important invention of a symbol for zero. With this symbol, 0, we are able to recognize the group to which a particular numeral belongs (that is hundreds, tens, or units, etc.) solely by its *position* in the number representation. Without the all-important zero we should not know whether 451 meant 4M5C1X, 4M5X1, or several other possibilities, but with it we can specify our meaning exactly. For example, when we write 4510 we know immediately that the meaning is 4M5C1X, so that the symbols for the groups are now entirely superfluous and can be removed. The importance of this is that now any number, *no matter how large,* can be written in its entirety with ten or fewer symbols. This is an enormous advance and , with a little effort, we can write numbers larger even than the number of grains of sand in the world, and only have to remember the ten numerals 0,1,2,3,4,5,6,7,8,9.

The development of the concept of zero, and its use in positional number systems, is attributed to the Hindus some twelve or thirteen centuries ago. The actual numerals which we use are commonly referred to as arabic numerals although their origin and development are not precisely known. The advantages of the positional system over the earlier counting methods are so great that it has become the closest thing which the world has to a universal language. This being the case, it is perhaps worth our while to examine it a little more closely.

The groups which we use are units, tens, hundreds, thousands, and so on, and can be conveniently written in a kind of shorthand notation by putting ten equal to 10^1, one hundred equal to 10^2, one thousand equal to 10^3, etc., where the number above and to the right of the ten is called a *power* (or an *exponent*) and indicates the number of zeroes coming after the one when the number is written out in full. This power or exponent can also be thought of as the number of tens which multiply together to give the larger number in question. With this terminology it is apparent that we can now easily write down extremely large numbers without wasting much ink. Moreover it also becomes clear that what we really understand by the number 14,658, for example, is one lot of ten thousands, four lots of one thousand, six lots of one hundred, five tens, and eight units, all added together or, in symbols

$$1 \times 10^4 + 4 \times 10^3 + 6 \times 10^2 + 5 \times 10^1 + 8.$$

More generally, for a number which contains $(n+1)$ digits, we write

$$a_n \ldots\ldots a_2 a_1 a_0,$$

where each subscripted digit a is a numeral and can take a value between 0 and 9, and we understand it to mean

$$a_n \times 10^n + \cdots + a_2 \times 10^2 + a_1 \times 10 + a_0.$$

It is now a very simple step from here to be able to write numbers and to count in bases (or groups) other than ten. For instance, if instead of 10 we wish to count in groups of 5, then the number $a_n \cdots a_2 a_1 a_0$ now means

$$a_n \times 5^n + \cdots + a_2 \times 5^2 + a_1 \times 5 + a_0,$$

but where each subscripted digit a can now take only one of the five values 0,1,2,3,4.

If the human race had evolved with only one five-fingered hand rather than two, then among the host of other inconveniences there might have appeared a counting system in groups of five such as that referred to above. To this race of one-handed people the number 14,658, for example, which seems so ordinary to us, would make no sense at all. To them it would look something like the number 3?7%2 appears to us. We should say that the symbols ? and % are not numerals and therefore don't mean anything in the context of numbers; they would say that the symbols 6, 5, and 8 are just as meaningless. Nevertheless, they are able to count just as well with their base-5, or 'quinary', system as we can with our decimal one. For the number of marks

$$* \ * \ * \ * \ * \ * \ * \ *$$
$$* \ * \ * \ * \ * \ * \ * \ *$$

they would write 31, understanding it as three times the base (that is five) plus one. We, on the other hand, write this same number as 16, understanding it as one times the base (that is ten) plus six. Note that the system used by the 'quinary people' has no separate symbol for five. This is their base and they would write it as 10 and, let us suppose, call it ten as we do. Pointing to each of the stars above in turn they would count as follows:

$$1, \ 2, \ 3, \ 4,10,11,12,13,$$
$$14,20,21,22,23,24,30,31.$$

If you use the fingers of one hand to help, and count out loud, one, two, three, four, ten, eleven, twelve, thirteen, fourteen, twenty, and so on, the new system soon becomes familiar at least for smaller numbers.

Thus, having fewer numerals than ten with which to count is no great incovenience. In fact, at first sight it might appear to be an asset, since it requires us to remember fewer numerals. Let us take this procedure of reducing the number of numerals to its limit to investigate the ultimate in supposed counting efficiency. The smallest number of numerals which can be used to make up a counting system is two. This is the *binary* system, and for it we need only two different symbols, a zero 0 and one other, say 1. That certainly seems simple enough. How would the counting go? Since there is no separate symbol for two, this would be our 10 and counting would start from 1 as follows: 1, 10, 11, 100, 101, 110, 111, 1000, 1001, 1010, 1011, 1100,1101, 1110, 1111, 10000,... in place of the decimal counting scheme: 1, 2, 3, 4, 5, 6, 7, 8, 9, 10, 11, 12, 13, 14, 15, 16, ..., and right away we can see a problem developing. The length of these base-two, or 'binary', numbers is increasing much faster than their decimal equivalents. Indeed, the binary equivalent of what we normally think of as one thousand is already quite formidable at 1,111,101,000, although it can be readily understood by interpreting it in the form

$$1 \times 2^9 + 1 \times 2^8 + 1 \times 2^7 + 1 \times 2^6 + 1 \times 2^5 + 1 \times 2^3$$

$$= 512 + 256 + 128 + 64 + 32 + 8 = 1,000.$$

For everyday purposes this rapid increase in the number of digits obviously leads to complexities which greatly outweigh the convenience of only having to remember two symbols 0 and 1. What, for example, should we call this number 1,111,101,000 if we wished to speak about it? Presumably it would be one billion, one hundred and eleven million, one hundred and one thousand; hardly a simplification over its decimal equivalent of 'one thousand'. At the other extreme we might think of trying a very large base such as one hundred. In this system the decimal number 'one hundred' would be our new 'ten'. The drawback, of course, is the fact that we should need to know and recognize no less than one hundred different symbols, which would be required for the numerals. It seems clear that for greatest efficiency, at least in human society, some compromise is desirable between the two conflicting conveniences of having few numerals (or symbols) and having short numbers. Looked at from this point of view, the decimal system with its base of ten does not seem to be too bad a choice. There are those who think that a duodecimal system, with a base of twelve, would be better, since the 'duodecimal 10' (that is twelve) would be exactly divisible by one, two, three, four, and six, whereas the 'decimal 10' is only exactly divisible by one, two, and five. However, since society is hardly likely to make such drastic changes at this late date, we can feel fortunate that the counting system literally *hand*ed down to us is, for our daily needs at least, close to one with optimum efficiency.

Before we leave this little excursion which we have made into counting, it is fun to note that it is also possible to count in a system for which the base is a negative integer. Although no evidence exists that any civilization has ever done so, the scheme in fact does have some advantages. A counting system of this kind can include all the minus numbers $-1, -2, -3, -4, -5, \ldots$ in addition to all the plus ones without the need to distinguish between them by introducing a special sign ($-$) to denote the numbers smaller than zero. There is nothing very special about how the new system works; it just follows the same rules that we set out for the positive based systems.

As an example we might consider counting to the base -10. What would we mean by the number 136 in this system? Well, following our earlier rules, which are general for any base, we must mean

$$1\times(-10)^2 + 3\times(-10) + 6$$

which, if we recall from our schooldays that a minus times a minus makes a plus, works out to be $100 - 30 + 6$ or 76 in our regular decimal number language. What about the number 1360? Obviously this means

$$1\times(-10)^3 + 3\times(-10)^2 + 6\times(-10)$$

and works out to be $-1000 + 300 - 60$, which is -760, in regular numbers. It follows that in the base -10 counting system the number written as 136 is a positive integer, while that written as 1360 is a negative one. How, you may ask, do the smallest counting numbers go in base -10? Well, counting up from one on our fingers, for example, would go like this:

$$1,2,3,4,5,6,7,8,9,190.$$

What is this one hundred an ninety which crops up so unexpectedly? It is simply ten in our new system as can easily be checked out:

$$190 = 1\times(-10)^2 + 9\times(-10) = 10.$$

Convinced of this it is now easy to check that, continuing to count on our toes, the next set of ten integers come out like

$$191,192,193,194,195,196,197,198,199,180.$$

But we could now count equally well starting from zero and going down the minus numbers. In our ordinary decimal system it is necessary to invent the new symbol '$-$' and proceed in the following manner:

$$-1,-2,-3,-4,-5,-6,-7,-8,-9,-10.$$

In the new system, however, the counting goes as follows;

$$19,18,17,16,15,14,13,12,11,10,29,28,27,....$$

as you may readily check out, and no new symbol is necessary at all.

The counting sequence in base -10 certainly looks a bit strange, but it is perfectly logical. It is true that there are some difficulties for the casual observer. How, for example, do you tell by looking at a base -10 number whether it is positive or negative? The answer is quite simple; it is positive if it contains an odd number of digits and negative if it contains an even number of digits. There are also simple rules for telling which of two numbers is the larger - another point which looks a bit baffling to the beginner - so that the system suffers mainly from lack of familiarity rather than any other shortcoming. New rules have to be learned for adding, subtracting, multiplying, and dividing, but they exist and are quite simple, although I do not intend to confuse you further by going into more details.

The reader will perhaps now feel a sense of relief to learn that the rest of this book will concentrate on numbers written only in our familiar everyday decimal notation. Nevertheless, it is worthwhile to stress just one more time that the fundamental properties of numbers which make them so fascinating (such as their possible primeness, for example) are quite independent of the language in which we choose to express them. Indeed, whenever we have recourse to appeal to the all-powerful modern-day computer to help us in our efforts, we are using a system which works in a base-two notation. Since the computer utilizes electronic switches, which can be either on or off, it is ideally suited to counting in that 'binary' system which we found to be so cumbersome for humans. Translating an 'on' switch as 1 and an 'off' switch as 0 it can go searching for prime numbers, for example, with great speed and efficiency. Provided that it translates its findings back into our common decimal system before printing them out, its output is immediately understandable and the results are just as valid, although they have been entirely calculated in the binary world, as if some mathematical 'superman' had managed the task working throughout with ordinary numbers.

2. THE SEARCH FOR PRIME NUMBERS

Of all the numbers which the modern-day mathematician works with, the simplest to get a feel for are the 'counting numbers' 1,2,3,4,... . These are more formally referred to as the *natural* numbers, or the *integers,* and it is tempting to think that no scientific subject could possibly be simpler to study or to understand. Surprisingly, there is ample evidence to suggest that precisely the opposite is true. The 'theory of numbers' (meaning the natural numbers) is both one of the oldest and most challenging of sciences, abounding with tantalizing conjectures and unproven assertions to this day. It is perhaps the greatest of all challenges to the power of pure mathematical reasoning and the greatest treasury of mathematical truths.

Every integer can be broken down into constituent parts in a variety of ways such that, in a sense, each has a distinct personality. In this manner some groups of integers fall rather naturally into 'families' which have a particular characteristic in common. The best known family of all, and one which has maintained a fascination and mystery for mathematicians for well over 2,500 years, is the prime number family. Prime numbers are integers which can be divided exactly (that is without a remainder) only by themselves and by 1. Thus, the smallest ones are

$$2, 3, 5, 7, 11, 13, 17, 19, 23, 29, 31, 37, ...$$

and they continue on to larger and larger values, becoming more and more tedious to calculate. It would be of enormous assistance in studying them if there were some sort of pattern in their appearance or if they had some outward sign to distinguish them from the rest. Unfortunately, if such a pattern exists or if such a sign be present, it has eluded discovery to this day. On the other hand this does not mean that a very great store of knowledge has not been accumulated over the centuries concerning what are colloquially referred to as 'the primes'.

One of the first questions which was asked by the early Greek mathematicians was 'do the prime numbers go on forever?' Since there are 15 primes between 1 and 50 (the number 1 itself is not normally counted as a prime) and only 10 primes between 50 and 100, it might appear that the primes become less densely distributed among the integers as we go to higher and higher values. A check of larger numbers seems to confirm this 'thinning out' (although the effect is rather slow and irregular) so that it is then only natural to ask whether the prime numbers might eventually stop altogether. This question was asked and answered by the early Greeks themselves, and it was the great Euclid who first provided the answer in about 300 B.C. He argued in an impressively simple way as follows: suppose for the moment that they do stop. In this case there must then be some largest prime number of all. Let us call it N. Now consider the much larger number made up of all the prime numbers, up to and including N itself, multiplied together. This number looks like

$$2 \times 3 \times 5 \times 7 \times 11 \times 13 \times 17 \times \cdots \times N$$

and evidently is exactly divisible by every prime which exists - if the original conjecture is true. Now let us add 1 to this very large number to make a new number. This new number cannot be exactly divided by any of the original primes up to and including N since, from the way it is formed, division by any one of them will now always leave a remainder of 1. Therefore our new number, which is far larger than N, is either a new prime number itself or it is divisible by a new prime number larger than N. In either case we have established that there is a prime number larger than the one N which was assumed to be the largest . It follows that no such largest prime number can exist and that the primes therefore do go on, like the integers, for ever. Having understood this Euclidean proof, it is now of interest to ask whether numbers of the form

$$(2 \times 3 \times 5 \times 7 \times \cdots \times N) + 1$$

do usually turn out to be primes themselves, or whether they tend to be not prime but divisible by a prime number larger than N. For the smallest values of N, namely the first five primes $N = 2, 3, 5, 7, 11$, these numbers (which for the record are 3, 7, 31, 211, and 2311 respectively) are all prime. Surprisingly, thereafter these numbers are virtually all *composite* (which is the fancy word mathematicians use for those numbers which are not prime). In fact, the only other prime numbers of the above form for N less than 1000 occur when $N = 31$ and $N = 379$. Readers who are interested in recent computer investigations of these and related types of numbers can turn to Appendix 1 at the back of the book where further details are given.

Mathematicians have been searching for centuries for a simple method which would enable them to determine, without great effort, whether or not a particular number is prime. It is true that methods do exist, but none materially shortens the work of testing from basic first principles, which implies checking the possible divisibility by all primes less than the square root of the number. Nevertheless, a general method does exist for listing all primes from the smallest without missing any. The procedure is extremely simple in principle, but unfortunately becomes unwieldy when the numbers get too large. It was first proposed about 250 B.C. by the Greek philosopher Eratosthenes and is usually referred to as the 'sieve of Eratosthenes'.

The method is so simple as to be almost obvious. It consists of writing down all the integers up to a predetermined limit of interest, and of eliminating all the numbers which are composite (or not prime). The sieve is started by knocking out all the even numbers (which are all divisible by the first prime number 2). Having done this, the smallest remaining number is the second prime 3. We now eliminate all the numbers divisible by 3 (which are called *multiples* of 3) from those which survived the first sifting operation. Five is now the first number remaining, so its multiples drop out next, then the multiples of 7, then of 11, and so on. If at each stage we note the number we are sifting by, and record it as a prime, we gradually build up a list of all the prime numbers from the smallest with none omitted. This basic principle is still used today in formulating programs for modern electronic computers in order to generate primes. The computers, of course, can go to very much larger numbers than Eratosthenes could manage, but only by virtue of their quickness of operation and not by any significant improvement in understanding the fundamental code which imbeds the primes throughout the number system.

It would be so much nicer if a formula could be derived to generate prime numbers; even if it wasn't able to give them all. Many suggestions have been forthcoming over the centuries, but none has been successful although some have come enticingly close, only to fail on closer inspection. Consider for example the simple formula

$$n^2 - n + 41,$$

in which n is to be put equal to the positive integers starting from 1. Trying n equal to 1, 2, 3, 4, ... , in turn reveals that this formula generates prime numbers all the way up to $n=40$. Alas, it fails for $n=41$. At this value the formula gives $41^2 - 41 + 41$, which is equal to the perfect square 41×41. The equally simple formula

$$n^2 - 79n + 1601$$

does even better. Once again putting n equal to 1, 2, 3, 4, ..., and so on, we find that it generates prime numbers all the way up to $n=79$

only to fail us at $n=80$.

If we learn anything from this experience, it is that the apparent success of a formula for a finite number of tries does not guarantee anything. We should really be looking for a general property of some kind, rather than grasping for straws. Unfortunately prime numbers are frustrating objects which seemingly grow like weeds, without any discernable pattern, in the boundless garden of natural numbers. We have listed all the prime numbers less than 1000 in Table 1 for you to examine in case, by chance, you should care to check things out for yourself!

TABLE 1
THE PRIME NUMBERS UP TO 1000
2,3,5,7,11,13,17,19,23,29,31,37,41,43,47,53,59,61,67,71,73,79, 83,89,97
101,103,107,109,113,127,131,137,139,149,151,157,163,167,173,179, 181,191,193,197,199
211,223,227,229,233,239,241,251,257,263,269,271,277,281,283,293
307,311,313,317,331,337,347,349,353,359,367,373,379,383,389,397
401,409,419,421,431,433,439,443,449,457,461,463,467,479,487,491, 499
503,509,521,523,541,547,557,563,569,571,577,587,593,599
601,607,613,617,619,631,641,643,647,653,659,661,673,677,683,691
701,709,719,727,733,739,743,751,757,761,769,773,787,797
809,811,821,823,827,829,839,853,857,859,863,877,881,883,887
907,911,919,929,937,941,947,953,967,971,977,983,991,997

But even if we know very little about the precise pattern in which the prime numbers occur, we do know something about algebraic expressions like those in the formulas above. Expressions of this type, which are made up of terms in n, n^2, (and possibly higher powers of n) plus a number, are termed *polynomials* by the mathematicians. A bit of thought on our part can easily establish that these types of formulas can never succeed in giving only primes. Thus, if we write a general polynomial in the form

$$a + b \times n + c \times n^2 + d \times n^3 + ...,$$

where a, b, c, and d, are integers, it is clear that when $n = a$, every term in the expression is exactly divisible by a so that the entire polynomial, which is the *sum* (that is addition) of these terms, must also be exactly divisible by a. This formula can therefore never generate a prime when n is equal to a no matter what we choose for b, c, d, etc., or how many terms we decide to put into it. We now see, without doing any calculation, that the first formula

$$n^2 - n + 41$$

used earlier was bound to fail when n reached 41, while the second formula

$$n^2 - 79n + 1601$$

was also bound to fail eventually when $n = 1601$ (for a now well-understood reason) even if it had not succumbed at $n = 80$ through sheer misfortune.

It follows that mathematics, although not providing us with the secret of the primes so far, does at least tell us where *not* to look for a suitable formula, and thereby does save us a lot of futile stumbling down useless pathways. All mathematical expressions are not polynomials, of course, so that there is still plenty of room for speculation. For example, one might notice a simple numeral pattern in the appearance of particular prime numbers themselves, and hope that this pattern will always generate prime numbers. Such an assertion may be extremely difficult to prove but, if it is not true, a little testing of some of the predicted numbers may soon spell out its demise. This, unfortunately, seems to be the common fate of such efforts to date. An example or two may shed light on this approach. One might notice from a table of prime numbers that 31 is prime, and that so also are 331, 3331, and 33331. This seems to be a promising beginning; is it perhaps a pattern which could give prime numbers indefinitely? There is certainly no known mathematical proof which says that this is impossible. The pattern has been pursued and does produce additional primes with 333,331, 3,333,331, and 33,333,331 but eventually, as always seems to be the case, it fails; this time at 333,333,331 which is exactly divisible by 17.

One further well documented effort in this search for a formula which always gives primes was made by the famous seventeenth century French mathematician Pierre de Fermat who proposed that the expression

$$2^{2^n} + 1$$

would only generate prime numbers when n was put equal to 1, 2, 3, 4, ... and so on. The first term in this formula means 2 raised to the power 2^n, and this number gets extremely large very quickly as we continue along the series $n=5,6,7,8$ etc. so that, until the dawn of the computer age, it was very tough to test any but the first few members of the series. Let us first consider the value of the exponent 2^n which is just n twos multiplied together. Starting from $n=1$ it goes like 2, 4, 8, 16, 32, etc. It follows that the predicted prime numbers, or Fermat 'primes' as they are commonly called, begin as

$$F_1 = 2^2 + 1 = 5$$
$$F_2 = 2^4 + 1 = 17$$
$$F_3 = 2^8 + 1 = 257$$
$$F_4 = 2^{16} + 1 = 65,537$$
$$F_5 = 2^{32} + 1 = 4,294,967,297$$

and, as we can see, very rapidly become quite formidable. In Fermat's day it was known only that the first four were indeed primes, but whether that fifth one with ten digits (let alone the higher terms of the series) was also prime had not been determined. About a century after Fermat proposed this sequence of 'primes' (and we must remember that Fermat never claimed to have proved that they were so) the Swiss mathematician Leonhard Euler, about whom we shall be hearing much more, established that the fifth Fermat 'prime' F_5 was not a prime number at all but was equal to

$$6,700,417 \times 641.$$

In the normal course of events this would have ended all interest in these Fermat 'primes' but, surprisingly, they popped up again in an entirely different context in the early nineteenth century. In addition, with the arrival of the electronic computer in recent years, it has now become possible to check many of the larger Fermat numbers for primeness. Incredibly, not a single prime number in the Fermat series has yet been found larger than F_4 although scores have been tested. The present-day question has therefore been reversed from that originally posed by Fermat and now is 'Do *any* Fermat numbers beyond $n=4$ yield prime numbers?' This question remains open and is the object of much numerical research because of the new relevance given to Fermat primes which is sketched out below.

The story of the manner in which the Fermat numbers reappeared as objects of attention in mathematics is particularly interesting, since it involves pure geometry and the early Greeks. The Greeks, in the time of Euclid and thereafter, were very much interested in finding methods for constructing regular geometric figures by using just a pair of compasses and a ruler. Now regular geometric figures are those which have all their sides of equal length and all their angles equal. The simplest example is an equilateral triangle and the next simplest is a square. The former has three equal sides, and three angles all equal to 60 degrees, while the latter has four equal sides and four angles all equal to 90 degrees (or a right-angle if you prefer). These regular figures, or polygons as they are known, exist with any number of sides, the next simplest being five-sided figures (pentagons), six-sided ones (hexagons), and seven-sided ones (heptagons). The Greeks were able to construct the regular triangle, square, and pentagon and, by repeatedly bisecting angles (a construction which could be carried out in the manner prescribed), they could also generate regular N-sided polygons where N was one of the values

$$N = 3,6,12,24,36,48,....$$
$$N = 4,8,16,32,48,...$$
$$N = 5,10,20,40,60,...$$

They were also aware of the fact that a 15-sided polygon could be constructed, but beyond that they were unable to make any further progress. For example, try as they would, they were not able to construct the regular heptagon (with $N=7$). There the matter rested for many centuries until 1796, when the young German mathematician Karl Friedrich Gauss was able to determine by algebra exactly which N-sided figures can be constructed with a pair of compasses and a ruler, and which cannot. His remarkable result was that regular polygons can be so constructed if, and only if, N is a power of 2, or is a power of 2 multiplied by one or more prime numbers of the Fermat form. Since mathematicians define any number to the power zero to be equal to 1, in order to conform with the general rules of algebra, the phrase 'power of 2' can be taken to include the number $2^0 = 1$. From this fact it also follows that the number 3 is really the smallest Fermat prime number, and should be written as

$$F_0 = 3,$$

since it is generated by the Fermat formula if we put $n = 0$. With this embellishment, the *only* odd-numbered regular polygons which can be constructed with rule and compasses are those with a number of sides equal to a Fermat prime, or to two or more Fermat primes multiplied together. These numbers begin 3, 5, 15, 17, ... and do not contain $N = 7$. In trying to construct the regular heptagon, the Greeks were therefore attempting the impossible. After the pentagon the next odd-numbered regular figure which can be constructed is one with 15 sides, and the manner in which a regular 15-sided figure could be constructed was known to the Greeks. Following the 15-sided figure, however, comes one with 17 sides, and this knowledge was new. Beyond the regular 17-sided figure are only two other constructable odd-numbered regular polygons with less than 100 sides. They are those with 51 (which is 3×17) and 85 (5×17) sides.

This new geometric association for the Fermat 'primes' created a fresh interest in these numbers. In particular, Gauss' rule for constructable regular polygons required Fermat numbers which truly were primes. Thus, for example, the regular polygon with $2^{32} + 1$ sides cannot be constructed with a rule and compasses by even the most diligent of geometricians since, although $2^{32} + 1$ is a Fermat number, it is known to be composite. As we mentioned earlier, in spite of intense activity using modern-day computing facilities, no Fermat number greater than F_4 (which is 65,537) has yet been found which is a prime number. Until such time as one is found, if indeed any more are ever found, the largest odd- numbered regular polygon which can in principle be constructed solely with a rule and compasses is one with

$$3×5×17×257×65,537 = 4,294,967,295$$

sides. If you find this rather mind-boggling, it is sobering to note that Gauss was only eighteen years old when he combined algebraic and geometric arguments to establish this as fact.

In the search for larger prime numbers F_n of the Fermat form, all values of n between 5 and 19 have been tested and found to be composite. At this writing, therefore, the smallest number which could possibly be a new Fermat prime is the Fermat number F_{20} or

$$2^{2^{20}} + 1.$$

Beyond $n = 20$ many larger ones have already been tested and all found to be composite. These include the numbers F_n with $n = 21, 23, 25, ...,$

52, 55, 58, 63, ..., 125, 144, 150, ..., 284, 316, 452, ..., 1,945, 2,023, 3,310, 4,724, 6,537, and others appear in the literature with each few passing months. Mathematicians now feel that there are at most only a finite number of Fermat primes and, of course, it is still quite possible that those for $n = 0,1,2,3$ and 4 are the only ones.

3. THE WORLD RECORD HOLDERS

Although we know that the number of primes is infinite, there is nevertheless some recognition to be gained among the prime number fraternity from discovering the largest known prime at any particular moment in time. Nowadays, with powerful computers at our command, the associated fame tends to be fleeting, but before the advent of the electronic computer the fame was more permanent, and records sometimes stood for very long periods of time. The race has been going on for centuries and many mathematicians have striven to be part of it. Euler, in particular, held the record for well over one hundred years in the pencil and paper era. More recently the frequency of record breaking has accelerated considerably. Today the record rarely stands for more than a few years and sometimes for only a few days; just long enough to digest the next computer output. By this we do not wish to imply that no challenge is left for the aspiring record breaker or that programming a computer for an attack on the largest prime number pinnacle is easy. With a given computer the heights to which one can aspire are still a function of the efficiency of one's method for checking for primeness, so that the challenge is certainly still there. On the other hand the research worker with the most powerful computer is certainly at an advantage. The complete list of prime number record holders since 1750, together with their respective primes, is shown in the accompanying Table 2.

From the table we notice that the record holders are almost exclusively prime numbers of the form $2^p - 1$, in which p also stands for a prime number. Numbers of this form are called Mersenne numbers and are named after the French mathematician Father Marin Mersenne. He first focused attention on them by making a statement in the year 1644 which implied that, for values of p less than or equal to 257, there were only 11 numbers of the form $2^p - 1$ which were prime. He was also bold enough to give those values of p for which (he claimed) these special numbers were prime; they were

$$p = 2, 3, 5, 7, 13, 17, 19, 31, 67, 127, 257.$$

Now this statement was not proved (and indeed is now known to be not quite correct) but it did involve many numbers which were larger than mathematicians of that time were able to handle, and it did set in motion an intense study of these 'Mersenne numbers' by amateur and professional mathematicians alike.

TABLE 2			
THE LARGEST KNOWN PRIME NUMBER			
Number	Discoverer	Year	Length in Digits
$2^{31}-1$	Euler	1750	10
$2^{127}-1$	Lucas	1876	39
$(2^{148}+1)/17$	Ferrier	1951	44
$180(2^{127}-1)^2+1$	Miller	1951	79
$2^{521}-1$	Lehmer	1952	157
$2^{607}-1$	Lehmer	1952	183
$2^{1279}-1$	Lehmer	1952	386
$2^{2203}-1$	Lehmer	1952	664
$2^{2281}-1$	Lehmer	1952	687
$2^{3217}-1$	Riesel	1957	969
$2^{4253}-1$	Hurwitz	1961	1281
$2^{4423}-1$	Hurwitz	1961	1332
$2^{9689}-1$	Gillies	1963	2917
$2^{9941}-1$	Gillies	1963	2993
$2^{11213}-1$	Gillies	1963	3376
$2^{19937}-1$	Tuckerman	1971	6002
$2^{21701}-1$	Nickel & Noll	1978	6533
$2^{23209}-1$	Nickel & Noll	1979	6987
$2^{44497}-1$	Slowinski	1979	13395

Numbers of the form 2^p-1 had appeared in earlier times, even in Euclid's day, in connection with the perfect numbers (which we shall meet later in Chapter 9). A little simple algebra shows that they cannot possibly be prime unless the exponent p is itself a prime number. The values of the Mersenne numbers up to $p = 19$ had already yielded seven primes in the years preceding Mersenne's pronouncement, but the situation for p in excess of 19 was quite unknown, and the manner in which Father Mersenne arrived at his conjecture is not known. The sixth and seventh Mersenne prime numbers had been given a generation or so earlier by the Italian mathematician Cataldi, and Mersenne himself never improved upon this, as far as we know, in the sense of actually proving that any of the larger numbers were truly prime. Indeed, in spite of all the activity following Mersenne's conjecture, the next prime number in the Mersenne series was not firmly established until the year 1750, when Euler showed that $2^{31}-1$

was a prime number just as Mersenne had suggested it would be. At that time this number, the eighth Mersenne prime, was the largest prime number known, and was to remain so for 126 years, until 1876, when the French mathematician Lucas achieved what then seemed to be an almost impossible feat. He established that the number $2^{127}-1$ was prime. Now this number, written out in full, is

$$170,141,183,460,469,231,731,687,303,715,884,105,727$$

and contains 39 digits; and its primeness was established with just pencil and paper. Incredible as this may seem, the fact that it was also another of Mersenne's predictions was equally astounding. It was almost as if Mersenne had received his information by Divine Revelation.

Lucas was able to accomplish his phenomenal feat by discovering new and much simpler methods of testing for primeness in numbers of the Mersenne form. It is these special methods, and their development over the years, which have perpetuated the emphasis on Mersenne primes in the race for the largest known prime number to this day. Lucas, however, had understandably not tested all the possible primes p between 31 and 127, so that in 1876 it was not known whether the new record holder $2^{127}-1$ was the ninth Mersenne prime in increasing order of magnitude, or whether there were others yet to be discovered between $p=31$ and $p=127$. In this range, in particular, was one more of Mersenne's predictions with $p=67$. Using Lucas' criterion for testing the Mersenne-type numbers, the missing values of p up to 127 were all eventually checked over the next few decades. In the year 1883, some 239 years after Mersenne's original pronouncement, came a most interesting development. The number $2^{61}-1$ was shown to be a prime and $2^{67}-1$ to be composite. Here was the first breakdown of the Mersenne predictions, although many felt even then that the $p=67$ in the Mersenne original list could well have been a misprint for $p=61$. A vestige of invincibility therefore still remained surrounding the assertion of the late French Father and persisted until the years 1911-1914, during which period the primeness of the numbers 2^p-1 with $p=89$ and $p=107$ was established, and the fact that there were indeed errors in the original Mersenne list finally confirmed.

None of these numbers, however, was larger than Lucas' record-holding prime $2^{127}-1$ which, with the completion of the checking of all values of p up to 127, was now firmly established as the twelfth Mersenne prime. So inevitably the search for the thirteenth Mersenne began in earnest. But it soon became apparent, as values of p larger than 127 were tested, that this next prime if it existed (since there was not - and still is not - any proof that primes of the Mersenne form go on forever) was at a much larger value of p than the frequency of occurrence of the earlier primes suggested (..., 61, 89, 107, 127,...?) Finally all values up to and including $p = 257$ (the last of the original Mersenne predictions) were checked out and no new prime number discovered.

Not until the dawn of the age of the electronic computer, after the Second World War, did Lucas' mighty number, written out above, finally lose its world-record-holding status. In 1951, after a reign of 75 years, it finally gave way to a computer-generated non-Mersenne number with 44 digits. But soon the Mersenne primes took over again as, with machine aid, the thirteenth member of the series (with $p = 521$) was finally located in 1952. This number contained more than four times as many digits as the next smaller Mersenne, and its discovery was achieved by Professor D. H. Lehmer who, using the National Bureau of Standards' Western Automatic computer, was the first to make a massive computer assault on prime numbers of the form $2^p - 1$. His efforts were crowned with considerable success, and during the year 1952 he found and published four more primes in the Mersenne series with exponents

$$p = 607, 1279, 2203, 2281.$$

The last number $2^{2281} - 1$, with 687 digits, made the 39 digit former record-holder of Lucas look puny indeed and was excellent testimony to the power of electronic computation. Nevertheless, computer development progressed dramatically during the nineteen fifties and sixties so that no record seemed safe for long. As can be seen from Table 2, all the subsequent record holders have been the Mersenne primes in increasing consecutive order. By 1971 the record was held by the twenty-fourth Mersenne prime with $p = 19,937$, a number with over 6000 digits. Since this number had nearly twice the number of digits as the twenty-third Mersenne, some suggested that further success using the same computer techniques might now be slow in coming.

Fortunately two fifteen-year-old high school students, Laura Nickel and Curt Noll of Hayward, Calfornia, were undaunted and in 1975 set out to hunt for the next Mersenne. Three years of effort were finally rewarded in 1978 when they announced the discovery of the twenty-fifth Mersenne prime with a p-value of 21,701. Adding their names to the list of world prime-number record-holders obviously only gave them further confidence and, after revising their computer program for the more efficient testing of even larger p-values, they discovered the twenty-sixth Mersenne prime, with $p = 23,209$, in early 1979. But in the computer age, alas, fame is fleeting. Only two months later (in April of 1979) a new and much more powerful computer, the Cray 1, was put to work on the Mersenne primes and David Slowinski of Cray Research Incorporated announced the discovery of the twenty-seventh Mersenne. This number

$$2^{44497} - 1$$

is the world record holder at the time of writing and has no less than 13,395 digits, almost twice as many as the twenty-sixth Mersenne. For the record it starts with 8545... and ends with ...8671. Using the latest generation of fast computing facilities all values of p up to 50,000 have now been covered.

There are therefore just 27 prime numbers of the form $2^p - 1$ with values of p less than fifty thousand. After all these centuries of effort to find them, it is now interesting to look at the exponent values p corresponding to these primes and to search for any discernable pattern. The prime-number p-values are

2, 3, 5, 7, 13, 17, 19, 31, 61, 89, 107, 127,

521, 607, 1279, 2203, 2281, 3217, 4253, 4423,

9689, 9941, 11213, 19937, 21701, 23209, 44497.

One might notice, for instance, that the smallest Mersenne primes

$$2^2 - 1 = 3$$
$$2^3 - 1 = 7$$
$$2^5 - 1 = 31$$
$$2^7 - 1 = 127$$

are all members of the p-value series above. Could this be a clue? Unfortunately, as is so often the case, it is nothing but sheer coincidence and the next member of the Mersenne series

$$2^{13} - 1 = 8,191$$

is not in the p-series. Numbers starting with 1 seem to be overly represented, but this is a more general phenomenon occurring in many groups of numbers and will be discussed in some detail in its wider context in Chapter 6. The phenomenon is not fully understood in all its ramifications, but it is certainly not related to the Mersenne p-numbers specifically. Quite generally, as one looks down the p-number series, it is seen that p increases anywhere from a few percent to more than a factor of four as we go from one member to the next; no even approximate degree of uniformity is apparent. It seems difficult to perceive anything of value; but maybe we can learn one thing. In a dissertation entitled *An Elementary Investigation of the Theory of Numbers* written by an early nineteenth century mathematician Barlow and published in the year 1811, one finds a statement that the largest prime number known is the number $p = 31$ of the Mersenne series. This is Euler's long standing record holder, and of it Barlow writes that it "is the greatest that will ever be discovered for, as they are merely curious without being useful, it is not likely that any person will attempt to find one beyond it". We learn perhaps that it is very easy indeed to underestimate the bounds of human curiosity.

4. THE DISTRIBUTION OF PRIMES

Although prime numbers of the Mersenne form are few and far between, we must remember that they are not representative of primes in general and have achieved special interest only because efficient methods for testing them for primeness have been devised. In this chapter we wish to consider *all* the known prime numbers in increasing order 2, 3, 5, 7, 11, 13, ... onward and ever upward. Is there any pattern or regularity, for example, in their appearance? We must first recognize that calculating all the primes, without missing one, is a far more arduous task that just looking for a particular large but specific prime number. As a result we shall not be concerned here with prime number data out to thousands of digits, as we were with the Mersennes. Nevertheless, modern computer facilities have made it possible to generate every prime up to values of more than a trillion (that is 10^{12}) so there are certainly plenty of raw data to maintain our interest. Two seemingly contradictory facts can be established from this data. Firstly, the primes appear to obey no other law than that of chance, meaning that it is impossible to predict in advance where the next one will be except by testing all the subsequent numbers in order. That much is rather discouraging and is, perhaps, what we have come to expect of prime numbers. Secondly, however, in complete contradiction, they exhibit a stunning regularity when looked at 'from a distance', and in this respect the laws governing their general behavior are obeyed with great precision.

To demonstrate the first claim, we note that prime numbers which are separated by only one even number are found to exist up to the largest numbers tested, and are believed to exist no matter how high you go. They are called *prime twins,* and have been located to values far in excess of the present limit to which all prime numbers have been computed. When I wrote the first draft of this chapter the largest known prime twins were

$$(1159142985 \times 2^{2304}) \pm 1.$$

These numbers have 703 digits, start with 4337... and end with ...17759 and ...17761 respectively. Unfortunately, research never stops long enough to complete a book and I now find that larger ones have been found, with at least one pair

$$(1024803780 \times 2^{3424}) \pm 1$$

having more than one thousand digits. The smallest prime twins are obviously (3,5), (5,7), (11,13), (17,19), ... , while others a little larger are, for example,

$$(10,006,427, \quad 10,006,429)$$

and

$$(1,000,000,009,649, \quad 1,000,000,009,651).$$

At the opposite extreme from prime twins it is easy to establish that consecutive prime numbers can be as far apart as you like. In other words, there can exist *arbitrarily* long sequences of numbers all of which are composite. This is proved by noting that when we have reached a number of the form

$$1\times2\times3\times4\times5\times \cdots \times n,$$

which mathematicians write in the shorthand notation $n!$ (and call 'factorial n'), then the next $n-1$ numbers which follow $n!+1$ are all composite. For example, $n!+2$ must be divisible by 2 (since $n!$ by definition is divisible by 2) and $n!+3$ must be divisible by 3, and so on all the way up to $n!+n$ which is divisible by n. Since we can choose n to be as large as we please, we can generate as many consecutive composites as we like. It follows that when we get up to extremely large numbers, the next prime can be as close as the next odd number or can be a very long way away indeed. There must therefore be enormous irregularities in the occurrence of prime numbers *when they are looked at in detail*. In practice, the biggest gap between consecutive primes which occurs below three trillion (3×10^{12}) is the 651 composites between

$$2,614,941,710,599 \text{ and } 2,614,941,711,251.$$

However, the occurrence of large prime gaps is quite rare although (looked at from another point of view) this gap of 651 occurs a lot sooner than might have been expected from the formula $652!+1$, which is a number in excess of 3×10^{1553}.

In spite of the great irregularities in the appearance of primes when examined in detail, one finds, by actual count, that each hundred from 1 to 1000 contains, respectively, the following number of primes:

$$25, 21, 16, 16, 17, 14, 16, 14, 15, 14,$$

while in the ten hundreds between 1,000,000 and 1,001,000 the corresponding numbers are

$$6, 10, 8, 8, 7, 7, 10, 5, 6, 8.$$

We note that the prime numbers gradually become rarer, but the fluctuations looked at in this way do not seem so large. What we are doing here is akin to standing back a bit from the numbers and looking at primes in groups rather than as individuals. Thus, although there are great irregularities in the distribution of the individual primes, when the large scale distribution is considered it appears fairly smooth. Let us look at this rather unexpected aspect of the prime number distribution in more detail since, at last, we may now be able to say something definite about the way the prime numbers appear - if only on a coarse scale.

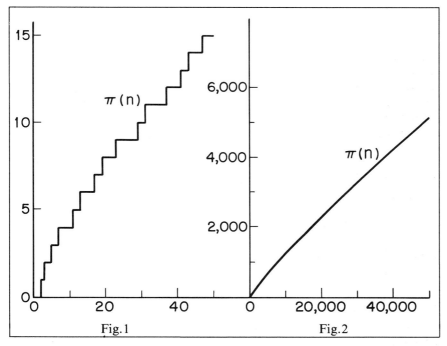

Fig. 1   Fig. 2

Examine, for a moment, the quantity which measures the number of primes smaller or equal to an arbitrary integer n. This quantity is a *function* of n (meaning that its value depends on n) and mathematicians like to write it as $\pi(n)$. Nevertheless $\pi(n)$, in spite of its fancy symbol, is just a number which begins at zero (when $n = 0$) and jumps up by one at every prime number it meets starting with 2, 3, 5, 7, etc. Consider this function $\pi(n)$ in the interval of numbers up to fifty. It is shown in Figure 1 and, in spite of the irregularities built into it, does seem to grow rather smoothly. What is amazing is the way this same function looks if we now plot it up to, say, fifty thousand (Figure 2). On the scale of our graph the fluctuations are now so small as to be invisible, and the regularity of the distribution of primes becomes extremely striking. What, one might ask, is the equation of this function? Can we perhaps write down some polynomial in powers of n represent it; or maybe some other kind of function? Karl Friedrich Gauss, perhaps the greatest mathematician of his time (whom we met earlier in Chapter 2), asked himself just this question when he was but fifteen years old. By studying the prime number tables, such as they existed in his day, and by extending them up to three million with calculations of his own, he noticed a regular pattern developing for the ratio $n/\pi(n)$, by which we mean n divided by $\pi(n)$, as follows:

n	$\pi(n)$	$n/\pi(n)$
1,000	168	6.0
10,000	1,229	8.1
100,000	9,592	10.4
1,000,000	78,498	12.7
10,000,000	664,579	15.0
100,000,000	5,761,455	17.4
1,000,000,000	50,847,534	19.7
10,000,000,000	455,052,512	22.0

where we have added several more powers of 10 in n made possible by the modern-day computerization of the problem. We see that the ratio of n to the function $\pi(n)$ increases regularly by about 2.3 when we go from one power of ten to the next. Mathematicians recognize this 2.3 behavior as connected with what is called the *natural logarithm* of ten. For practical purposes the natural logarithm of a number n can be defined as the ordinary logarithm from a set of log tables multiplied by 2.303. It is usually written as $\ln(n)$, and for the moment we can just think of it as a number from a set of tables, although we shall meet it again later in the book (Chapter 19) and discover a little more about it. Gauss therefore observed that the ratio $n/\pi(n)$ behaves more and more like $\ln(n)$ as we go to larger and larger numbers n. This result, stated originally by the young Gauss in the year 1792, is called the *prime-number theorem,* and was finally proved rigorously many years later. Its importance is that it enables us to *predict* the number of primes less than some large number *n* with considerable accuracy without the need to find them all and count them explicitly.

The result does not mean that the difference between $\pi(n)$ and the function $n/\ln(n)$ is necessarily small in an absolute sense, but only that it is small compared to the number $\pi(n)$ itself. In more recent years more complicated functions have been found which approximate the actual prime number distribution even more closely. One of them which is not too complicated in its definition, but is still one of the best, also goes back to Gauss. It is called the *integral logarithm* and is usually written as $Li(n)$. For those readers who know a little calculus (and the rest of you need not worry at all about it) the formal definition of this function is

$$Li(n) = \int_2^n dx/ln(x).$$

The following table indicates the accuracy of $Li(n)$ when compared with the actual count of primes $\pi(n)$ less than or equal to n:

n	$\pi(n)$	$Li(n)$
100,000	9,592	9,630
1,000,000	78,498	78,628
10,000,000	664,579	664,918
100,000,000	5,761,455	5,762,209
1,000,000,000	50,847,534	50,849,235

More sophisticated functions can do even better but for us $Li(n)$ is quite sufficient and, if it were plotted in Figure 2, the difference between it and the real $\pi(n)$ could not be seen on the scale of that figure.

All this is very impressive, but we must not be deceived by it. Our original statement concerning the randomness of prime numbers at the final digit level remains all too true. For example, from the above table the function $Li(n)$ tells us to expect six or seven primes per hundred in the immediate vicinity of n equal to ten million. This is not much help in deciding which numbers between 10,000,000 and 10,000,100 are actually prime. Surprisingly enough, there are only two prime numbers in this interval, namely 10,000,019 and 10,000,079, while there are no less than nine primes in the hundred integers preceding ten million, including the prime twins 9,999,971 and 9,999,973. In addition, if we look at the difference between the function $Li(n)$ and the actual number of primes $\pi(n)$, using the table above, it seems as if the approximation $Li(n)$ is always a bit too large. Not only this, but the amount by which it is too large seems to get bigger as n increases. For example, we find overestimates of 38, 130, 339, 754, and 1701 when n reaches the values 10^5, 10^6, 10^7, 10^8, and 10^9 respectively. Now it would hardly take a betting man to wager heavily that $Li(n)$ is always going to be bigger than $\pi(n)$ given this evidence. Unfortunately for him, outrageous as it may seem, he would lose. You see it has been established theoretically that, although the difference between $Li(n)$ and $\pi(n)$ does continue to increase for ever *on the average,* the fluctuations about this average can become so great that for some extremely large values of n the number of primes must exceed the estimate given by $Li(n)$. To date no such numbers have actually been found, but it has been proved that they exist and, moreover, that there must be at least one of them which is smaller than

$$10^{10^{10^{34}}}.$$

The sheer enormity of this number is difficult to appreciate. It has been called the largest number which has ever served a useful purpose in mathematics, and mathematical numbers always exceed by far those used in the other sciences. For example, the total number of atoms in the entire universe is 'only' about 10^{75}. This last number, written out in full, is a 1 followed by 75 zeroes. The monstrous number above, on the other hand, has no less than

$$10^{10^{34}}$$

zeroes. This means that, even if we used as many zeroes as there are atoms in the universe, we still could not come close to writing it out in full. This seems to bode ill for anyone seeking to find the *actual* value of n for which the number of primes first becomes larger than its theoretical estimate $Li(n)$. Be this as it may, the example does show us how unwise it can be to jump to conclusions on the basis of numerical data out to a puny few billion.

Another example of this kind of thing is given by a fascinating quantity which we shall label $p(n)$. It counts the number of *different* prime numbers which must be multiplied together to make up any number n of interest. For example, the number $n = 40$ can be expressed in the form $2 \times 2 \times 2 \times 5$ and has 2 such primes (or prime *factors* as they should more properly be called) namely 2 and 5, making $p(40)$ equal to 2. Now, according to a fundamental theorem of arithmetic, every positive integer can be made up as a *product* (that is a multiplication) of prime numbers in just one way, so that the number $p(n)$ is quite uniquely defined. Let us write down a few more examples:

$$6 = 2 \times 3$$
$$30 = 2 \times 3 \times 5$$
$$210 = 2 \times 3 \times 5 \times 7$$
$$625 = 5 \times 5 \times 5 \times 5.$$

This tells us immediately that $p(6)=2$, $p(30)=3$, $p(210)=4$, and $p(625)=1$. What sort of function is this $p(n)$? It is evident from the few examples already given that it certainly does not increase steadily with increasing values of n. Moreover, since the prime numbers go on forever, and since $p(n)$ is equal to 1 for all prime numbers, there must be some values of $p(n)$ equal to 1 no matter how large n becomes. On the other hand, the value of $p(n)$ for other numbers can grow much larger as n increases. Indeed $p(n)$ can become extremely large for some n-values because, if we think back to that number which we called factorial n and wrote $n!$ earlier in the chapter, it contains by its definition *all* the prime numbers up to n as different prime factors. It follows that $p(n!)$ increases steadily towards infinity as n gets ever bigger. Nevertheless, even in favorable circumstances $p(n)$ is a number which grows rather slowly, even if it can get to arbitrarily large values eventually. In fact the numbers 6, 30, and 210, set out above with $p(6)=2$, $p(30)=3$, and $p(210)=4$, are the first to occur with $p(n)$ equal to 2, 3, and 4 respectively. We do not find a single integer with $p(n)=10$ until we get all the way up to $n = 6,469,693,230$. Thus, $p(n)$ is a very jagged function which increases extremely slowly on the average, but continues to get larger and larger forever while all the time (and infinitely often) dipping back all the way to 1. It is perhaps the last word in fluctuation concepts. Nevertheless, we can inquire about the way it behaves *on the average* without getting into problems concerning its extremely jagged form; and even its average behavior is fascinating.

If you believe that a snail is slow then compare this for explosive progress. At $n=100$ the average value of $p(n)$ is just 1.71, by which we mean that the average number smaller than 100 has rather less than two different prime factors. At $n=100,000,000$ it has struggled ahead to a value of 2.9. Again this means that the typical number smaller than one hundred million still has less than three different prime factors, a statement which most people find utterly beyond belief. But given this fact, what are the chances of the average of $p(n)$ ever reaching to say a million? Quite slim you might think! Yet remarkably it can be shown that this average does get forever larger and, given patience in counting, one can eventually arrive at an average value of $p(n)$ just a large as one wishes. But how long does it take? At the present rate of progress obviously a very considerable time. Let us take a nice healthy jump in n, say all the way to 1 with a hundred zeroes after it. This number 10^{100} has been called a googol by some, and is sometimes displayed as an example of a ludicrously large number - something to laugh at but much too large to be taken seriously. Somewhat disappointingly we find that the average value of $p(n)$ for n equal to a googol is just 5.4. In desperation we try n equal to 1 followed by not one hundred but ten billion zeroes. This is truly a number to be reckoned with but, even at this lofty height (which makes a googol seem utterly insignificant by comparison), the average value of $p(n)$ is just 23.9. What number then can we possibly choose to take this sluggish $p(n)$-average to the heights of which we know it is capable. Why, that number

$$10^{10^{10^{34}}}$$

of course, which we met just a short while ago. It is a measure of the magnificence of this number that, when used for n, it at last brings out the potential in the average value of $p(n)$ which we knew all along must be there - giving it a value in excess of 2×10^{34}.

5. PRIME RACES, EMIRPS, AND MORE

If we look at the prime numbers up to 1000 given in Chapter 2, we find that they can be divided into separate groups in more than one way. For example, they can all be obtained by adding fours either to the number 3 or to the number 5. Split up in this way we can form two separate series of prime numbers as follows:

$$3, \quad 7, \ 11, \ 19, \ 23, \ 31, \ 43, \ 47, \ 59, \ 67, \ 71, \ 79, \ 83, \ \dots$$
$$5, \ 13, \ 17, \ 29, \ 37, \ 41, \ 53, \ 61, \ 73, \ 89, \ 97, \ \dots$$

There is nothing mysterious about the ability to separate primes in this way; it follows very simply from the fact that every prime number (except for the first prime 2 which we ignore) is odd, and that every odd number must be one more or one less than a number of groups of four (or what mathematicians call a multiple of four).

If we start each row, in the manner shown above, with the numbers 3 and 5 and write down all the primes in turn in increasing order, putting them into the row to which they belong according to the 'adding fours' rule, then at any particular moment either the top or the bottom row is longer (that is, contains more numbers). We can consider this as a number 'horse-race' to infinity. In the sequences set out explicitly above we have included all the prime numbers up to 100 (considering 100 as the first 'furlong post' in this number race which goes on forever). At this point the top or '3' row is ahead by two numbers. Does the '3' row stay ahead for long, or perhaps forever? I will not spoil the fun by telling you the answer at the $N = 1000$ marker - you can place your wagers and then find out for yourself by using the prime number table in Chapter 2. As far as the infinite limit is concerned, it has been shown theoretically that each row leads infinitely often so that the race itself can never be said to be won outright by either side. At any finite 'distance post', on the other hand, the race does have a leader (or is tied) and the competition is keen. It happens that this particular race is rather unusual and has been pursued all the way up to prime numbers of the order 2×10^{10}. Those interested in the details will find them in Appendix 2 at the back of the book.

The 'groups of four' number race set out above is not the only race in the prime number stakes by any means. For example, all the primes larger than 5 can be grouped by sixes. Starting at 7 in one row and at 11 in the other, adding sixes also separates the prime numbers into two groups as follows:

$$7, 13, 19, 31, 37, 43, 61, 67, 73, 79, 97, \ldots$$
$$11, 17, 23, 29, 41, 47, 53, 59, 71, 83, 89, \ldots$$

At the $N = 100$ mark in this race the contestants are neck and neck. Again, pick your favorite, use the prime number table, and check out the race up to $N = 1000$.

If a 'two horse' race is too tame for you, then races with more number-competitors are equally simple to set up. All prime numbers larger than 5 can also be derived by starting respectively with 7, 11, 13, and 17, and adding eights. In this manner we obtain four rows as follows:

$$7, 23, 31, 47, 71, 79, \ldots$$
$$11, 19, 43, 59, 67, 83, \ldots$$
$$13, 29, 37, 53, 61, \ldots$$
$$17, 41, 73, 89, 97, \ldots$$

Once again the competition at the $N = 100$ marker could hardly be more fierce, with the '7' and '11' rows dead level and just one number ahead of the '13' and '17' rows. What is more, since it is the opinion of most number theorists that each 'horse' in all number races of this kind gets ahead infinitely often in the long run, the wager on the leader at any particular 'number post' is a fair one without the necessity of giving the 'odds'. Another 'four-horse' number-race can be obtained for primes greater than 5 by noting that all such prime numbers end in a 1, 3, 7, or 9. We can therefore set up a '1-ending' prime number row, a '3-ending' row, a '7-ending' row, and a '9-ending' row as follows:

$$11, 31, 41, 61, 71, \ldots$$
$$13, 23, 43, 53, 73, 83, \ldots$$
$$7, 17, 37, 47, 67, 97, \ldots$$
$$19, 29, 59, 79, 89, \ldots$$

This is really a 'grouping by tens' arrangement exactly similar in general form to the others. Again, as you can see, the rows are very evenly matched.

The setting up of these races in no way implies that we know anything about prime numbers in general. They are just formulated to cover all the odd numbers except for sets which are obviously composite and therefore contain no primes at all. Thus, for example, in the 'grouping by sixes' we are omitting the series of odd numbers which starts with 9 and adds sixes, since these are all obviously divisible by 3. In this same spirit it is possible to set up many more prime number races with even more 'horses', and it should not be difficult for you to invent some more for yourself.

In this wider context another interesting thought occurs. If the number row containing your particular favorite is running a little behind the other, you will probably be interested in learning 'how fast the horse can go and for how long the burst of speed can be maintained'. The maximum speed is evidently defined by consecutive successes (that is primes) in the series, so that the important question concerns how many consecutive prime numbers you can (given the luck of the Irish) hope to find in any particular race. Some answers and some speculations concerning this question have been given by mathematicians. It is now known that in races like the ones above, which are formed by adding groups of n, it is not possible to get more consecutive primes than one less than the smallest prime number m which will not divide n exactly. As an example, in the four horse race above where we group by eights (that is in which $n = 8$), the value of m is 3 so that we can never expect to get more than two consecutive primes in any one of the four rows. In this race it is obvious that bursts of speed are not sustained at all, so that overtaking the leader is likely to be a slow and tedious process if one falls too far behind. In the two-horse race with $n = 6$, on the other hand, the situation is not quite so bad since the smallest prime not dividing 6 exactly is $m = 5$. It follows that, in this race, we can expect occasional bursts of 4 consecutive primes. As we can see, both horses in that race experienced just such bursts of speed early in the race; at 11, 17, 23, 29 and at 41, 47, 53, 59 for the '11' row. and at 61, 67, 73, 79 for the '7' row.

In order to get truly impressive bursts of sustained speed it is unfortunately necessary to form sequences in which the constant number to be added is quite large. If this number is 210, for example, then the smallest prime number not dividing 210 exactly is 11, so that we might expect 10 consecutive primes of this kind to arise from time to time. Such a group of ten consecutive successes is known, namely

199, 409, 619, 829, 1039, 1249, 1459, 1669, 1879, 2089

but runs of this length are far from easy to locate.

Before we leave prime number races, one interesting point should be recorded about the series

$$5, 13, 17, 29, 37, 41, \ldots$$

which was involved in the very first contest set out in this chapter. It was Fermat who first proposed, and Euler who eventually proved, that all the primes in this particular series, but not in its competing partner, can be represented in just one way as the sum of two squares. We illustrate this below by giving the explicit form for the members up to 100:

$$5 = 2^2 + 1^2 \qquad 13 = 3^2 + 2^2 \qquad 17 = 4^2 + 1^2$$

$$29 = 5^2 + 2^2 \qquad 37 = 6^2 + 1^2 \qquad 41 = 5^2 + 4^2$$

$$53 = 7^2 + 2^2 \qquad 61 = 6^2 + 5^2 \qquad 73 = 8^2 + 3^2$$

$$89 = 8^2 + 5^2 \qquad 97 = 9^2 + 4^2.$$

This was noticed originally by directly testing all the smaller prime numbers in this series and it is interesting, and perhaps comforting, to note that not all patterns concerning prime numbers which are suggested by the behavior of the first several terms of a series, necessarily turn out to be misleading. There is always the chance, as with this particular case, that the pattern is not accidental but is an indication of some general law. Whether, on the other hand, that law is easily proved is quite another question.

Finally we should note that in setting up prime number races of a more general kind (that is with a more complicated rule than just adding even numbers to a series of low primes) there is no proof in most cases that the primes necessarily go on forever within the series. Thus, for example, we may choose to set up races between prime numbers of the form n^2+1, n^2+2, and n^2+3. The first series starts off with $n=1$ (leading to the prime $1^2+1=2$), followed by $n=2$ $(2^2+1=5)$ and $n=4$ $(4^2+1=17)$, while the second and third series start with $n=1,3,9$ (leading to the primes $n^2+2 = 3,11,83$) and $n=2, 4, 8$ (leading to $n^2+3 = 7,19,67$) respectively. With these rules this prime number race up to one thousand looks like

n^2+1 :   2,   5,   17,   37, 101, 197, 257, 401, 577, 677, ...
n^2+2 :   3,   11,   83, 227, 443, ...
n^2+3 :   7,   19,   67, 103, 199, 487, 787, ...

with the n^2+1 'horse' leading comfortably. Just as much fun can be had with races of this kind as with the others, except that now the larger primes tend to be fewer and farther between and the mathematicians do not claim to know anything definite about the outcome. Whether one series actually leads in the infinite limit (and by this we mean that there is some 'milepost' N beyond which one row stays ahead for ever), or whether one or more of the series eventually stop altogether with an absolute highest prime number of the required form, is just not known. In other words, for this and like cases the race may have a true winner or, on the other hand, one or even all of the number 'horses' may fail to complete the course. Unfortunately, the final outcome for any particular case is not likely to be determined by using a pocket calculator and a table of prime numbers. Nevertheless we can still study the progress of the race in its early stages and speculate on a leader at the various 'mileposts' available to us. Prime number races therefore come in an enormous variety of possible forms and are really limited only by imagination, patience, and access to a suitable prime number list.

THE EMIRPS

Among the boundless number of primes can be found many interesting subgroups. There are, for example, some primes which read the same both forward and backward. These, like the numbers 151 and 727, are called number *palindromes*. A somewhat larger group are the primes which turn into other different primes when reversed. Since the word 'emirp' is prime spelled backwards, these numbers have been called the *emirps*. By their definition they obviously occur in pairs, and the smallest ones can easily be found from the table of primes given in Chapter 2; they are (13, 31), (17, 71), (37, 73), (79, 97), ... The ones given are the only two-digit emirps which exist in base-ten but there are many more with a larger number of digits. There are, in fact, 13 pairs of three-digit emirps, 102 pairs of four-digit emirps, and no less than 684 pairs of five-digit emirps.

Since the number of emirps seems to be increasing at a nice healthy rate as the number of digits gets larger, it is tempting to guess that they exist up to prime numbers of arbitrarily large decimal form. On the other hand, as we have seen before, judging the trends of number sequences from their behavior through the first few powers of ten can be hazardous. Just how hazardous in the present context can be seen by considering that particular subgroup of emirps which have no repeating digit. Quite evidently, from the above examples, all the two-digit emirps fall into this category. It happens that 11 of the 13 three-digit emirps are also of the non-repeating-digit type, as are 42 of the four-digit emirps and 193 of the five-digit emirps. This series

$$4, 11, 42, 193, \ldots$$

also seems to be increasing impressively. Yet by its very definition we know that this last series must fall to zero by the time it reaches the eleventh term, because there are only ten different numerals 0 through 9 and it becomes impossible to make up any eleven-digit number (prime or otherwise) without using at least one of them twice over. In fact, with a little additional thought, it is possible to establish that the non-repeating-digit emirps must fall to zero by the tenth term, since all ten-digit non-repeating numbers must contain each of the numerals 0 through 9 once only. They must therefore all have a sum of digits equal to

$$1 + 2 + 3 + \cdots + 8 + 9 = 45$$

which is exactly divisible by 9. Now there is a simple law of arithmetic which states that any number with a sum of digits which is divisible by 9 is itself divisible by 9. It follows that all ten-digit numbers of this kind are divisible by 9. None of them is therefore prime, emirp or otherwise. This means that the series above, which starts out so confidently with 4, 11, 42, 193, ... must come all the way back down to zero by the tenth term. The exact manner in which this series comes back to zero is unfortunately not known at the time of writing, since no-one has yet counted non-repeating-digit emirps up to nine digits.

In the more general context of all the emirps (repeating-digit and non-repeating-digit) there seems to be no simple argument concerning their possible demise as there was for non-repeating-digit ones. No-one has counted them out far enough even to allow for an educated guess.

Therefore we do not yet know whether the general emirps go on forever or whether they too eventually end with a largest emirp of all. We must all wait for the computers to catch up with our inquisitiveness. In the mean time there is ample opportunity for another friendly wager, for those who have a firm opinion on the likely outcome and who still have anything left to wager after all the prime number races.

Before leaving the emirps we might mention two of them which are very special. In listing the emirps up to ten million, two have appeared which are completely *cyclic*. By this we mean that the process of moving the first digit to the rear, and repeating this operation as often as you please, generates only other emirps. The smaller of these two 'cyclic emirps' is the five-digit number 11,939. By the cycling process described above it generates the additional emirps 19,391, 93,911, 39,119, and 91,193. The larger 'cyclic emirp' is the six-digit number 193,939. It produces the additional cycled numbers 939,391, 393,919, 939,193, 391,939, and 919,393, emirps all! Some purists might argue that one of the five-digit cyclic forms of 11,939 (namely 19,391) is not a true emirp because of its completely reversible (that is palindromic) form, but we shall leave that point for the reader to decide upon for himself. Of the six-digit numbers generated by 193,939, none is palindromic, so that a rigorous 'emirpness' is maintained throughout. Thus, if we are purists and insist that a true emirp must generate a *different* prime number when reversed, then there is only one known 'cyclic emirp', namely 193,939. We may now speculate whether it is unique among all decimal numbers, or whether out there in the countless number of primes lies another waiting to be discovered.

THE PRIME PRIMES

Consider for a moment the prime number 73,939,133. There is something quite unusual about it. By dropping successively the right-most digit, the remaining numbers are still all primes. Specifically, the successive prime numbers which we can generate in this manner starting from 73,939,133 are

73,939,133   7,393,913   739,391   73,939

7,393   739   73   7

and, because of this property, they are sometimes called *prime primes*. The largest of the particular sequence of prime primes (that is the number 73,939,133 in the example above) is called the *generator* of these numbers. Now the generator given above is not alone among the primes although such generators are known to be rather few in number. It is unusual in that it is the very largest prime prime generator which exists in the decimal system; not just the largest known at present, but the largest one there is. Moreover the complete list of such generators is small enough that we can give them all in a table of rather modest size. Specifically they are:

53	7,331	373,393	7,393,933
317	23,333	593,993	23,399,339
599	23,339	719,333	29,399,999
797	31,193	739,397	37,337,999
2,393	31,379	739,399	59,393,339
3,793	37,397	2,399,333	73,939,133
3,797	73,331	7,393,931.	

When trying to extend this list from eight to nine digits, it is found that there just are no nine digit prime generators at all. It therefore follows that there can be no larger generator, and no larger prime prime, than 73,939,133 among the entire infinite number of primes.

There is, of course, nothing special about paring down the prime generators from the right-hand-side. One could equally well set up a list of equivalent prime primes by truncating numbers from the left, that is by removing the left-most digit and repeating the process. It turns out that this list is a very much longer one than that shown above for right-pared generators, although it is still of finite size. We shall give here only the largest generator of left-hand-side prime primes. It has no less than 24 digits and is

357,686,312,646,216,567,629,137.

It therefore generates a sequence of twenty four prime numbers which ends with 9,137, 137, 37, and 7.

A FEW PRIMES WITH INTERESTING DIGIT PATTERNS

Among the very large prime numbers perhaps the simplest to remember are those which contain a repeating pattern of some kind. Easiest of all, I suppose, are those which contain only a single repeating digit, or those in which the digits appear in their 'correct' order 1, 2, 3, 4, ... etc. It is obvious that any 'single-repeating-digit' number, like 7,777,777 for example, cannot possibly be prime because it is exactly divisible by the digit itself (in this case 7). This is true unless, of course, the digit in question happens to be a 1. The number 1,111,111, for example, is not obviously composite. Are there any prime numbers of this sort? This question is easily answered because the very smallest example of such a number, namely 11, is itself a prime. But are there any more? Well, 111 is not a prime since it is 3×37, and 1,111 is also not a prime, being 11×101.

The question of whether any number larger than 11, comprised only of a string of ones, is prime is a very interesting one and remained unanswered for many years; until the 1960's in fact. These numbers, made up of *rep*eating *'units'* or 'ones' are sometimes called *repunits* from an obvious abbreviation. Since there is a simple law of arithmetic which states that any number made up of digits adding up to a number divisible by 3 is itself divisible by 3, it follows immediately that one can rule out numbers made up of 3, 6, 9, 12, 15, ... 'ones', since they all fall into this category. It is also easy to show that any number made up of an even number of ones is always exactly divisible by 11. Together these rules remove a large fraction of the repunits to be tested but, for the ones which remain, there are usually no simple tests to come to our aid. Nevertheless, over the years all the smaller repeating-one numbers were checked and found to be composite, so that it soon became apparent that if there was another prime of this kind (that is in addition to 11) then it was quite a formidable number. But there is such a prime and it was eventually located with the help of the electronic computer. It is

$$1,111,111,111,111,111,111$$

and has nineteen ones. It is now known that there are others as well, but they are of the utmost rarity. In checking numbers of up to one thousand repeating ones, only two others have been located. They have respectively 23 and 317 ones. Needless to say, it is not known whether prime numbers of this form exist to arbitrarily large values.

Related to the repunit prime numbers are those which are all of this kind except for the first or last digit. Some examples of prime numbers of this form are

$$111,111,113$$
$$11,111,111,113$$
$$11,111,117$$
$$11,111,119$$

the last two making a rather remarkable pair of prime twins. Other examples, with the differing digit at the front of the number, are

$$61,111,111$$
$$71,111,111$$
$$31,111,111,111 \quad .$$

Of the prime numbers which have the numerals in their 'proper' order 1, 2, 3, 4, ... etc., two interesting examples are

$$1234567891$$

$$123456789123456789123456789 1$$

Must these numbers necessarily end in a 1? Prime numbers larger than 5 can normally end in any of the numerals 1, 3, 7, or 9. For these 'cyclic primes' starting with 1234...., however, we can easily rule out endings by 3 or 9 since the total sum of the digits in numbers like

$$123456789123 \ ..... \ 89123$$

and

$$123456789123 \ ..... \ 89$$

are always divisible by 3, so that the numbers themselves must always be divisible by 3 as well, no matter how many cycles are involved. This leaves us with 1 and 7 as the only possible end-digits for prime numbers of this kind. Is there a good example of one ending with a 7? Indeed there is, the longest yet discovered in fact. It begins, as required, with 1234.... and cycles seven times before finishing with 1234567.

6. THE BAFFLING LAW OF BENFORD

The origin of the unusual phenomenon to be discussed in this chapter is said to go back to an observation, commonplace before the era of the pocket calculator, that log tables in public libraries appeared to be dirtier at the beginning than at the end. Since the users of tables of logarithms do not usually read them like a novel, starting on page one, this can hardly be explained by the fact that these tables of numbers lack the appeal to grip the reader who therefore gives up before reaching the end. Presumably it must imply that the numbers which science or engineering students had to deal with tended to start with the lower numerals like 1 and 2 rather than with the higher ones like 8 and 9.

A first reaction to this, I think, is that such a notion is preposterous. Numbers of interest to scientists, or anyone else for that matter, should surely (in the decimal system) be just as likely to start with any one of the nine numerals between 1 and 9. What would these numbers be anyway? Presumably they would be largely numbers taken from scientific and statistical tables. Is there really any evidence that numbers taken from such lists favor the lower numerals and if so, why? This question, absurd as it first seemed, was finally taken seriously by the physicist Frank Benford in the year 1938. He investigated twenty such tables, ranging from the surface areas of lakes and rivers to the molecular weights of thousands of chemical compounds. His conclusion, after classifying more than twenty thousand entries according to their first digits (regardless of the position of the decimal point), was both surprising and fascinating. It is shown in Figure 3 where the frequency of first digits is indicated in the nine divisions from 1 to 9.

Benford found that the numeral 1 appeared as the first digit in a number in his list with a frequency of 0.306. By this he meant that no less than 30.6% (or nearly one third) of the numbers started with a 1. The frequency for the larger numerals fell off in a regular fashion, as can be seen from the figure, until less than 5% were found to start with an 8 or with a 9. If each of the nine available numerals had appeared equally often, then this frequency of appearance would have been 0.111 (or a little over 10%) for each. Students using these figures would therefore get the first page of their log tables fully six times as dirty as the last.

Does this peculiar distribution of first digits persist in other lists? Not wishing to accept on trust any results which I could easily verify for

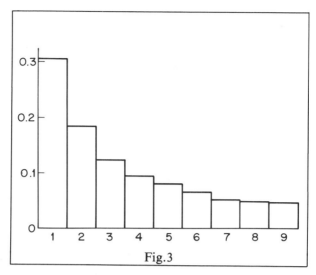

Fig.3

myself, I turned to the Marquis edition of *Who's Who in America* and checked out the street numbers in the home addresses of the first several hundred biographees. Even after as few as 500 addresses were examined the first digit phenomenon suggested by Benford's work was again quite clear. In this sample I found a little over 28% of the street numbers beginning with a 1, falling to various values between 4% and 6% for the numbers beginning with 6, 7, 8, and 9. What can it mean?

First of all it is necessary to define the effect a bit more precisely. Quite obviously not *all* tables of numbers behave in this way. If, for example, we look at a set of log tables themselves (which record the logarithms of all numbers between 1000 and 9999) then we find that this particular collection of numbers is quite different, being overly weighted at the 9 end of the spectrum. Also it is clear that if the list contains properties with only a small scatter about their average, such as for instance the annual rainfall where you live, then the first digit distribution will be very different from Benford's finding. The effect which we wish to pursue seems to be present in lists which contain a certain randomness (that is, lists for which we cannot even approximately predict the next member from some known general behavior) and which covers a wide range (typically several factors of ten) of numbers. Good examples might be street numbers, city populations, lengths of rivers, and areas of states or countries. Systematic tables like logarithms and square roots are likely to behave quite differently. As a result Benford, in his original investigation of this topic, offered the finding as a general 'law of *anomalous* numbers', implying by the word anomalous those numbers between which there is no known relationship. He even proposed an equation for this 'law' by noting that the distribution in Figure 3 is very closely given by

$$P(n) = log[(n+1)/n]$$

in which $P(n)$ is the probability that the first digit will be n, and log stands for the logarithm to the base ten, which is the normal logarithm found in log tables.

At this point it is perhaps desirable to say a word or two about what base-ten logs really are, although it is not necessary for the reader to know anything about logarithms in order to enjoy the Benford story. Those who are not eager to learn more at this time can skip the rest of this paragraph with the knowledge that painless logs will sneak up on us anyway in a later chapter in another guise. By $log(n)$ we mean the power to which 10 must be raised to get the number n. For example, $log(10) = 1$ since the power to which 10 must be raised to get 10 is just 1. Similarly $log(100) = 2$, $log(1000) = 3$, and so on. For values of n which are not convenient integer powers of 10 it is necessary to know what is meant by a fractional power. Quite generally $10^{m/n}$ means the nth root of 10^m, or equivalently, the number which when multiplied together n times makes 10^m. As a simple example $10^{1/2}$ is the number which when multiplied by itself gives 10. This is more commonly called the square root of 10 and is about 3.162. The log of 3.162 is therefore 1/2. Using this definition any number a can be expressed in the form 10^b where b is called the logarithm of a and is written $log(a)$.

From Benford's law above, the exact numerical probabilities $P(1)$ through $P(9)$ for finding an 'anomalous' number starting with a numeral 1, 2, 3, ..., 9 work out to be (to three decimal place accuracy)

$$P(1) = 0.301 \quad P(2) = 0.176 \quad P(3) = 0.125$$
$$P(4) = 0.097 \quad P(5) = 0.079 \quad P(6) = 0.067$$
$$P(7) = 0.058 \quad P(8) = 0.051 \quad P(9) = 0.046$$

Benford decided that this was nothing less than a law of nature, but others following him have been more sceptical. In particular, it now appears that not all of the collections of numbers which seem to obey the Benford distribution of first digits fall into the same category. Some of them, at least, can now be understood in a less mysterious way.

TABLE 3					
The probability $p(n)$ that a randomly chosen house number in a street with n houses consecutively numbered from 1 to n should have a first digit 1.					
n	$p(n)$	n	$p(n)$	n	$p(n)$
1	1/1	21	11/21	154	66/154
2	1/2	22	11/22		
3	1/3	23	11/23	197	109/197
4	1/4			198	110/198
5	1/5	50	11/50	199	111/199
6	1/6	51	11/51	200	111/200
7	1/7	52	11/52	201	111/201
8	1/8			202	111/202
9	1/9	97	11/97		
10	2/10	98	11/98	997	111/997
11	3/11	99	11/99	998	111/998
12	4/12	100	12/100	999	111/999
13	5/13	101	13/101	1000	112/1000
14	6/14	102	14/102	1001	113/1001
15	7/15	103	15/103	1002	114/1002
16	8/16			1003	115/1003
17	9/17	150	62/150		
18	10/18	151	63/151	1999	1111/1999
19	11/19	152	64/152	2000	1111/2000
20	11/20	153	65/153	2001	1111/2001

Consider, for example, the street numbers which I laboriously extracted from Who's Who in America. Suppose that I live in a street with 99 houses in it numbered consecutively from 1 to 99. Unfortunately American streets are seldom that predictably organized even if they do contain 99 houses, but that is another matter. If I am in such a street and choose a house number at random, then it is apparent that the chances of picking one with any particular numeral as its first digit (excluding 0 of course) is 11 out of 99, since there are 11 numbers starting with each numeral between 1 and 99. The corresponding probability is 11/99 or about 0.111. But what happens if I live in a street with only 19 houses numbered consecutively from 1 to 19. Clearly my chances of picking at random a house number starting with a 2 through 9 are now only 1 in 19. My chances of picking a house number starting with a one, on the other hand, are now 11 out of 19 or 0.579, which is more than 50%. More generally, in a street containing n consecutively numbered houses between 1 and n, the chances of randomly picking out a house number with the first digit 1 can easily be worked out by counting the house numbers which begin with a 1 and then dividing by n. These probabilities, as a function of n, are shown in Table 3 .

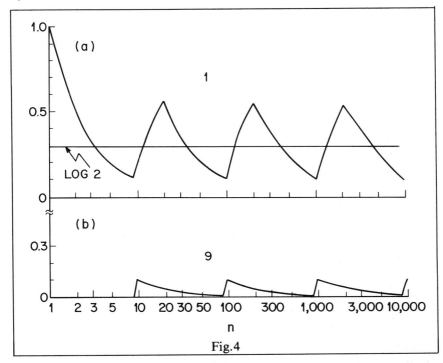

Fig.4

Between $n=1$ and $n=9$ they fall from 1 to a low of 1/9, then between $n=10$ and $n=19$ they rise up to 11/19. Beyond $n=20$ they fall steadily back to 1/9 (or 11/99 which is the same thing) which is reached wnen $n=99$, before rising again to 111/199 at $n=199$. This perpetual rising and falling behavior continues indefinitely. It is shown in Figure 4a for values of n up to 10,000. The curve has a sawtooth shape with peaks near 0.55 and valleys near 0.11. One thing is immediately clear; only if every street were to contain 9, 99, 999, 9,999 houses (etc.) would the chances of locating a house number at random with a first digit equal to 1 be as small as the naively anticipated value of 1/9.

We also show, in Table 4, the equivalent probabilities for finding a house number starting with the numeral 9. As a function of n it also has a sawtooth shape and is sketched in Figure 4b. Unlike its numeral-1 counterpart it never exceeds the value 1/9 (or 0.111) and

TABLE 4					
The probability $p(n)$ that a randomly chosen house number in a street with n houses consecutively numbered from 1 to n should have a first digit 9.					
n	$p(n)$	n	$p(n)$	n	$p(n)$
1	0	93	5/93	901	13/901
2	0	94	6/94	902	14/902
3	0	95	7/95	903	15/903
4	0	96	8/96		
5	0	97	9/97	950	62/950
6	0	98	10/98	951	63/951
7	0	99	11/99	952	64/952
8	0	100	11/100		
9	1/9	101	11/101	997	109/997
10	1/10	102	11/102	998	110/998
11	1/11	103	11/103	999	111/999
12	1/12			1000	111/1000
13	1/13	500	11/500	1001	111/1001
		501	11/501	1002	111/1002
87	1/87	502	11/502		
88	1/88			8998	111/8998
89	1/89	897	11/897	8999	111/8999
90	2/90	898	11/898	9000	112/9000
91	3/91	899	11/899	9001	113/9001
92	4/92	900	12/900	9002	114/9002

falls in the valleys down to values only one tenth as large (or 0.0111). Again it is clear that, except for streets with 9, 99, 999, 9,999, ... houses, the random chance of finding a house number with first digit 9 is not 1/9 but is always smaller than 1/9.

Now if we suppose that the chances of finding a street with n houses in it is about the same for any value of n then we can estimate mathematically the average chance of randomly locating a 'first digit 1' and 'first digit 9' house number when all houses in *all* streets are included in the sample. All we need to do is to draw a horizontal line through each of the two curves in Figures 4a and 4b such that half the sawtooth curve is above and half below this line in each case. Without going into any technicalities concerning the precise rules for averaging, it is clear from the figures that these lines will fall quite close to 0.3 and 0.05 for the numeral-1 and numeral-9 cases respectively. These values, if we look back to Benford's probability list, are very close to those given by his logarithmic law and found in practice. Although we recognize that all street lengths n are not equally likely (in fact they probably follow a Benford distribution law themselves), and most (at least in America) do not have numbers running consecutively from 1 upward, the proper 'real-life' averaging process must be closely linked to the one we have given so long as house numbers up to large values (in the hundreds and the thousands) actually occur with an acceptable frequency. In this sense, at least, much of the mystery of Benford's law of anomalous numbers has now been removed from my house number findings from *Who's Who in America* although the exact equation $P(n) = log[(n+1)/n]$ must be a rather idealized form for this case. This is all very well, but what about the other sets of numbers which obey the Benford relationship; the city populations, the areas of countries, the lengths of rivers (or streets), and so forth? A similar argument does not hold for these. Why do they also obey Benford's law? This time the answer is not so certain, although some progress has been made.

The first clue came from Benford's original tables themselves. Since some of his collections of numbers were of groupings involving lengths and areas, it was interesting to ask which set of units had been used for making up the tables. Thus, for example, areas could be measured in square miles, acres, square yards, or metrically in hectares, and each unit would provide a completely different set of numbers from the same raw data. Surprisingly it was found that the units used were completely irrelevant. In other words, the law of anomalous numbers was produced by Benford's lists no matter what units of length or area were used. All the individual numbers changed, of course, when a conversion from one unit to another was made, but the anomalous distribution of first digits persisted. This was a new and very significant finding.

Following this discovery the problem was taken up by Professor Pinkham of Rutgers University in the State of New Jersey. He was able to prove a very significant result which is mathematically exact. It is that any set of numbers which is *scale invariant* (that is which reproduces the same frequency of the nine first-digit numerals when they are all multiplied by the same arbitrary number) must necessarily have exactly the frequencies corresponding to Benford's logarithmic law. The actual proof is quite a sophisticated argument in probability theory and need not concern us, but the result is quite fascinating. It means that if you write down a set of numbers at random, and plot the frequency of occurrence of their first digits, and then multiply each of them by say 2 or 3 and plot the equivalent frequency for the new numbers, then the two frequency curves will in general be quite different. However, if you have chosen a set of numbers with the Benford distribution of first digits, then this distribution will be maintained after each member of the set has been multiplied by the same number regardless of what this number might be. Even with comparatively small sets of well-scattered numbers it is not difficult to verify the effect, although the larger the set the more accurate the findings. We demonstrate this 'scale invariant' phenomenon in Table 5 in which a list of 20 numbers is given which obeys the Benford distribution criterion about as well as such a small list can. Each number in the original list is then multiplied in turn by 2, 3, 4, 5, and 6 in order to generate five further lists. From Table 5 we see that the five generated lists reproduce the original distribution very closely with no error greater than one in any first-digit category. If we take the average of all the distributions it is found to come closer to the exact Benford distribution than is possible with a single 20-number set. To two significant figures the exact Benford distribution for a set of 20 is

$$6.0, \ 3.5, \ 2.5, \ 1.9, \ 1.6, \ 1.3, \ 1.2, \ 1.0, \ 0.9$$

and the averaged distribution from Table 5 does not differ from this by more than 0.3 for any digit.

Any readers who feel that these numbers have been rigged to produce the effect are welcome to choose their own sets of original numbers and their own sets of multipliers. The more well spaced the chosen numbers are, then the more impressive the scale invariance will be for a relatively small and manageable set. It is also interesting to take another set of numbers which does not approximately satisfy the Benford distribution and to verify that such a list has no scale invariance at all.

Pinkham's proof therefore seems to establish that nature somehow provides us with scale invariant lists. But why? Can we proceed any further? Well, it is interesting to note that Benford's distribution can also be exactly generated by forming a list in which each term is obtained by multiplying the previous one by the same number. For example, the two series

$$1, \ 2, \ 4, \ 8, \ 16, \ 32, \ \ldots$$

and

$$1, \ 3, \ 9, \ 27, \ 81, \ 243, \ \ldots$$

TABLE 5					
A set of numbers roughly obeying Benford's distribution of first digits is multiplied in turn by 2, 3, 4, 5, and 6. The first digit distribution for each generated set is shown at the bottom and can be seen to closely reproduce the original.					
×1	×2	×3	×4	×5	×6
10	20	30	40	50	60
12	24	36	48	60	72
14	28	42	56	70	84
15	30	45	60	75	90
17	34	51	68	85	102
19	38	57	76	95	114
22	44	66	88	110	132
25	50	75	100	125	150
28	56	84	112	140	168
32	64	96	128	160	192
35	70	105	140	175	210
38	76	114	152	190	228
43	86	129	172	215	258
47	94	141	188	235	282
53	106	159	212	265	318
57	114	171	228	285	342
61	122	183	244	305	366
73	146	219	292	365	438
85	170	255	340	425	510
97	194	291	388	485	582
1: 6	6	7	7	6	6
2: 3	3	3	4	4	4
3: 3	3	2	2	2	3
4: 2	1	2	2	2	1
5: 2	2	2	1	1	2
6: 1	1	1	2	1	1
7: 1	2	1	1	2	1
8: 1	1	1	1	1	1
9: 1	1	1	0	1	1

which involve successive multiplications by 2 and by 3 respectively, are of such a form and their 'Benfordian' behavior can be established using surprisingly few terms. When continued to 100 terms, the upper series gives the first digit distribution

30, 17, 13, 10, 7, 7, 6, 5, 5

while the lower gives

29, 19, 12, 8, 8, 7, 7, 5, 5.

These sorts of series are called *geometric sequences*. Could it therefore be that nature tends to count geometrically, that is by powers like 2, 4, 8, 16, rather than by 1, 2, 3, 4?

This in essence is the suggestion which Benford himself put forward when he first commented on the entire phenomenon. It implies that in a list of 'anomalous numbers', generated by some count of natural phenomena which covers a wide range, there is a tendency for the step between successive numbers (when arranged in ascending order) to be a fixed fraction of the preceding one. It also implies that in a wide distribution there will most likely be roughly the same number of events in each order of magnitude, at least near the center of the distribution where most of the events are.

To verify this I have checked my travel guide (now alas some years out of date, but that should be of no material consequence in the present context) which lists the populations of countries deemed to be of interest to potential tourists. The population list does follow Benford's first digit distribution quite well, and in addition contains (to within a factor of two) equal numbers of examples in the population categories $10^4 - 10^5$, $10^5 - 10^6$, and $10^6 - 10^7$. Thus, for this case at least, the list is indeed an example of counting geometrically rather than arithmetically. But is this a choice of the individuals who composed the list rather than a natural law of populations? Did the compilers of the travel guide want to give me a fair selection of countries of small, medium, and large populations so as not to appear biased? I cannot say with confidence. In fact, among mathematicians and statisticians there is still considerable controversy concerning the meaning and significance of the law of anomalous numbers. Does nature really provide us with scale invariant tables; or is it the person compiling the table who unconsciously performs this feat. I leave further research up to the reader.

7. WHAT IS SO SPECIAL ABOUT 6174?

What could possibly be special about a number like 6174; or perhaps we should say about 495 and 6174, because these two numbers are, in their own particular way, probably unique among all the decimal integers. Here we use the term 'decimal integers' because we are now going to look at digit patterns in numbers and these, of course, are dependent on the base to which we count. It is important to distinguish between the properties of numbers and of representations. The former (e.g. whether they are prime, perfect, amicable etc.) are absolute, while the latter depend on the counting base. Nevertheless, as we have seen with Benford in the previous chapter, a lot of fun can be had with representations as well as absolutes, and in this (and several of the following) chapters we shall look at the decimal representation in some detail. Whether as decimal representations, or as numbers in general, 495 and 6174 certainly do not exhibit any outward appearance of undue mystique; one is odd and one is even; they are not prime or perfect. In fact they give every indication of being rather ordinary, and in order to probe their particular brand of uniqueness it is necessary to go back to India and to the 1940's when the Indian mathematician D. R. Kaprekar started to play with integers in a new and rather fascinating way. It was really a type of solitaire game in the beginning, and the rules are simple enough for anyone to play. In the era of the pocket calculator the procedure is not even particularly time consuming, and the moves of the game are as follows:

a) Think of an integer which contains more than one digit.

b) Arrange the digits in decreasing numerical order so as to make the largest number possible out of them (not all the digits must necessarily be different but they should not *all* be the same).

c) Arrange the digits a second time; this time in increasing numerical order so as to make the smallest possible number out of them (putting zeroes in the front if necessary).

d) Now subtract the smaller number (c) from the larger one (b) to give a new number with the same number of digits; that is including initial zeroes if necessary.

e) Finally, repeat the operations (b), (c), and (d) on the new number and continue.

The object of the exercise is to see what happens as you go on and on. Let us try a very simple example to begin with; say a 2-digit number like 90. For this particular case the above procedure takes the form:

$$90 - 09 = 81$$
$$81 - 18 = 63$$
$$63 - 36 = 27$$
$$72 - 27 = 45$$
$$54 - 45 = 09$$
$$90 - 09 = 81$$
.....

and we notice that the last line is exactly the same as the first, so that we have formed a loop which regenerates itself over and over again. Does the same sort of thing happen if we try another 2-digit number? Well, you can test it for yourself, but it does soon become apparent that they all do degenerate into this same endless loop.

Suppose we now progress to 3-digit numbers; does anything different in principle happen for these? Let us take another particular example and try it out. How about the number 900? In this case the procedure progresses as follows:

$$900 - 009 = 891$$
$$981 - 189 = 792$$
$$972 - 279 = 693$$
$$963 - 369 = 594$$
$$954 - 459 = 495$$
$$954 - 459 = 495$$
.....

Aha! Here comes 495, the first of our special numbers. But this time it is not part of a loop; it stands on its own as the end product of the entire procedure - that is, it is self-reproducing forever. But what if we choose another 3-digit number? Try it for yourself and see! Was it 495 again? If so you may be tempted to ask 'does this *always* happen and do I have to test *every* 3-digit number in order to find out?' Mercifully the answer to the second question is no; as is often the case, a little bit of thought helps.

Consider a *general* 3-digit integer abc, where a, b, and c each represents a numeral between 0 and 9, assuming that they are not all the same. This number abc therefore stands for a quantity made up of 'a' lots of 100, 'b' lots of 10, and 'c' units, as follows

$$100a + 10b + c.$$

Now let us suppose that this number has already been arranged with its digits a, b, and c in decreasing order. The rules of the game now require that we subtract from abc the number in which the digits are reversed. This number cba in its expanded form stands for

$$100c + 10b + a$$

so that the required subtraction of rule (d) of the game now looks like

100a	+	10b	+	c
100c	+	10b	+	a
100(a-c)	+	0	+	(c-a)

The bottom line, which is the result of the subtraction, adds up to $99a-99c$ or, equivalently, $99(a-c)$ which is 99 lots of $(a-c)$. Since a is bigger than c (from our earlier assumption that abc is bigger than cba) and both a and c are numerals between 0 and 9, it follows that $(a-c)$ can take on only values between 1 and 9. The implication is that after only one subtraction routine, starting from any 3-digit number, we can get only one of nine possible results namely

$$99\times1, 99\times2, 99\times3, ..., 99\times8, 99\times9.$$

These numbers are

$$099, 198, 297, 396, 495, 594, 693, 792, 891.$$

After these have been arranged with their digits in decreasing numerical order to start the second subtraction routine, we are left with only five numbers (990, 981, 972, 963, and 954) which need to be tested further. It is now a simple process to check out these five in detail - in fact, all except 990 already appear in our example above - and confirm the unique property of 495 as the self-reproducing endpoint for all 3-digit numbers.

By now we are beginning to have some suspicion of what is special about 6174; but let us first check it out with an example. Try the 4-digit number 2359. With the now familiar rules we find the sequence

$$9532 - 2359 = 7173$$
$$7731 - 1377 = 6354$$
$$6543 - 3456 = 3087$$
$$8730 - 0378 = 8352$$
$$8532 - 2358 = 6174$$
$$7641 - 1467 = 6174$$

....

which finishes up with the number 6174 reproducing itself forever. It can be shown that this same self-reproducing number 6174 is obtained using *any* 4-digit starting number (so long as all the digits are not the same). Again it is not necessary to actually check out all 4-digit starting numbers to prove this, but the proof is just a little bit longer than it was for the 3-digit equivalent. In analogy with the general 3-digit case set out above, we now choose a general 4-digit number abcd and perform the first subtraction stage of the game as follows:

1000a	+	100b	+	10c	+	d	
1000d	+	100c	+	10b	+	a	
1000(a−d)	+	100(b−c)	−	10(b−c)	−	(a−d)	

The bottom line can be reassembled in the form $999(a-d) + 90(b-c)$ by simply adding together the terms. Now, since $a-d$ can be anything between 1 and 9 and (since b can possibly be equal to c) $b-c$ can be anything between 0 and 9, this leaves 90 numbers. After rearranging them in order of decreasing digit size (in order to begin the second subtraction routine) we find that only 30 separate forms remain to be tested. This is certainly not an impossible task, and confirmation of the uniqueness of the regenerating endpoint 6174 follows without too much effort.

Having dealt so successfully with 2-digit, 3-digit, and 4-digit numbers it is now only natural to seek the patterns for 5-digit numbers. As a starting point we shall once again choose an arbitrary example; say 87542. This generates the following sequence

$$87542 - 24578 = 62964$$
$$96642 - 24669 = 71973$$
$$97731 - 13779 = 83952$$
$$98532 - 23589 = 74943$$
$$97443 - 34479 = 62964$$
$$96642 - 24669 = 71973$$

....

in which the last line is identical to the second. It follows that we have generated a loop and that therefore no unique self-generating number can possibly exist for 5-digit integers. Of course, this single example does not establish that a self-reproducing number may not exist in addition to this loop (nor that there may not be other loops) but at least we already know that, unlike the 3-digit and 4-digit cases, loops do exist in the 5-digit hierarchy. Considerable computer work has been done on this 5-digit 'Kaprekar-problem' and it is now known that all 5-digit series end in one of three possible loops. No self-reproducing endpoint is obtained for any 5-digit series. The three loops obtained are the one in the example above, namely

$$71973, 83952, 74943, 62964,$$
another four-number loop
$$75933, 63954, 61974, 82962,$$
and the two-number loop
$$59994 \text{ and } 53955.$$

Although I have not seen a complete analysis for six or higher digit numbers (the complete investigation rapidly becomes more tedious and time consuming as the number of digits increases) several loops have already been found for numbers with 6, 7, 8, and 9-digits. Since the number of integers to be tested grows rapidly with increasing size, it seems highly unlikely that any complete group of larger numbers will ever reduce to a single self-reproducing endpoint. If this is true, it leaves 495 and 6174 in a very special position and leads us to wonder how many other rather ordinary looking integers might be unique in their own special way - if only we learn to ask the right questions.

Before passing on to other interests it should be stressed once again that the above procedure of Kaprekar, which makes 495 and 6174 so unusual, is only valid if we are counting in the decimal system. There is nothing special about 6174 cows or 495 dogs. It is only the base-ten representation of these numbers which has led us to focus attention on them. Using the same set of rules, we could play the game equally well counting in any base and come up with different loops etc. for 2, 3, 4, and more digit numbers in these new representations. Nevertheless, since I promised in my opening chapter that I would not stress non-decimal numbers in the rest of this book, I will quickly pass over this additional dimension of Kaprekar's game, leaving the adventurous enthusiast to take it up for himself if he so desires.

Before taking leave of Mr. Kaprekar we might, with interest, make note of another class of numbers which he discovered. In the year 1949 Kaprekar first introduced what he called the *self-numbers*, and these are best explained by starting with the basic procedure which (surprisingly) fails to generate them. This curious statement implies that there is a simple rule which can be used to generate most (but not all) integers, and the ones which cannot be so obtained are Kaprekar's self-numbers.

The basic method used to give all except the self-numbers is called *digit addition* and works as follows. Select any positive integer and add to it the sum of its digits. Take 58 as an example. The sum of 5 and 8 is 13, so that the number generated in this way is 58 + 13, which is 71. The procedure can be continued indefinitely, and in the specific case of 58 goes as follows

58, 71, 79, 95, 109, 119, 130, 134, 142, 149, 163, 173, ...

each number being generated by the number before it. The interesting question is whether all numbers have generators - that is, can they all be produced in this manner? Obviously from what I have implied earlier they do not, and it is almost trivial to verify that the single digit numbers 1, 3, 5, 7, and 9 have no generators. Therefore, at least for the smaller integers, self numbers (or numbers without generators) do not seem to be great rarities. Nevertheless, as we proceed to larger integers they do become a little scarcer, although they exist to arbitrarily high values. There are, in fact, only 8 self-numbers with 2-digits. They are

20, 31, 42, 53, 64, 75, 86, and 97.

We note that each is just 11 larger than its predecessor. Is this how the larger self-numbers are obtained in general? Not quite! Although most self-numbers do follow this pattern, occasionally there comes a 'break in the pattern' which adds interest to what otherwise would be a rather dull exercise. Thus, for example, the self-number series with two digits can be extended into the realm of three digits when we find the continuation

108, 110, 121, 132, 143, 154, 165, 176, 187, 198,

209, <u>211</u>, 222, 233, 244, 255, 266, 277, 288, 299,

where the numbers at the 'breaks' are underlined. Moreover the 'break in the pattern' is not always a jump of 2 (as it is in the above example) so that the series is not as regular as it first appears.

Kaprekar himself discovered the *general* method for testing a number to see whether or not it is 'self-born', to use his words. The method is most simply explained by giving an example. Let us first think of a rather large number, say 3,333,333,333 (for no particular reason) and ask whether it is a self-number. Kaprekar's method of testing proceeds as follows. First find the sum of all the separate digits which make up the number (in our case 30) and then add again the digits of this new number (to get 3). If this number is odd, as it is for our particular case, add 9 and divide by 2 (to get 6). If it is even simply divide by 2. Now subtract the number obtained by this strange recipe (6 for our case) from the number being tested, and check whether this gives a generator of the number. For our example we obtain 3,333,333,327 which is not a generator of 3,333,333,333 since its digits add up to 33 and

$$3,333,333,327 + 33 \neq 3,333,333,333$$

where the symbol $\neq$ means 'is not equal to'. The method of testing now proceeds by subtracting nines, each time checking to see if we have hit upon a generator, and we continue the process until we have checked for as many possible generators as there are digits in the original number. For our example we must check nine further numbers as follows:

$$3,333,333,318 + 33 = 3,333,333,351$$
$$3,333,333,309 + 33 = 3,333,333,342$$
$$3,333,333,300 + 24 = 3,333,333,324$$
$$3,333,333,291 + 33 = 3,333,333,324$$
$$3,333,333,282 + 33 = 3,333,333,315$$
$$3,333,333,273 + 33 = 3,333,333,306$$
$$3,333,333,264 + 33 = 3,333,333,297$$
$$3,333,333,255 + 33 = 3,333,333,288$$
$$3,333,333,246 + 33 = 3,333,333,279$$

Since 3,333,333,333 nowhere appears on the right hand side of these equations, it follows that we have not found a generator and that our particular example is indeed a self-number. What about 3,333,333,334;

is this a self number too? Well, add the digits (31); add them again (4); the result is even so divide it by 2 to get 2. Subtract 2 from the number to be tested (3,333,333,332) and check it as a possible generator. It is not. Now begin subtracting nines and try again:

$$3,333,333,323 + 29 = 3,333,333,352$$
$$3,333,333,314 + 29 = 3,333,333,343$$
$$3,333,333,305 + 29 = 3,333,333,334$$

where finally our test number has shown up on the right hand side. Therefore 3,333,333,334 is not a self-number but is generated by 3,333,333,305.

We notice something else from the test-series for 3,333,333,333 given above. The number 3,333,333,324 appears twice on the right hand side. This means that it has more than one generator. Thus, whereas some numbers have no generators at all (the self numbers), others can have two or perhaps even more generators. The smallest number with two generators turns out to be 101, which is produced by both

$$91 + 10 = 101$$
$$100 + 1 = 101.$$

How high do we have to go before we find a number with three generators? This is an interesting question and has a rather surprising answer; we have to go all the way up to 10,000,000,000,001. This rather monstrous number is generated by

$$9,999,999,999,892 + 109 = 1,000,000,000,001$$
$$9,999,999,999,901 + 100 = 1,000,000,000,001$$
$$10,000,000,000,000 + 1 = 1,000,000,000,001$$

Eventually, as we move on to even larger numbers, we encounter examples with 4, 5, and even more generators, but the sheer size of the numbers involved makes it very difficult to say much about them. Let it suffice for us to give here the smallest number with 4 generators; it is

$$1,000,000,000,000,000,000,000,102$$

and is generated by

1,000,000,000,000,000,000,000,100	+	2
1,000,000,000,000,000,000,000,091	+	11
999,999,999,999,999,999,999,902	+	200
999,999,999,999,999,999,999,893	+	209

As a final amusing observation we consider the powers of 10, that is 10^1, 10^2, 10^3, 10^4, ... and so on, to see whether any is a self-number. We observe that

5	+	5	=	10
86	+	14	=	100
977	+	23	=	1,000
9,968	+	32	=	10,000
99,959	+	41	=	100,000

but find that 1,000,000 (one million) has no generator and is therefore a self-number. Why then is a millionaire such an important person? Because, answers Kaprekar, the number one million is the first power of ten to be a self number. And how far do we have to go to find the next smallest power of ten which is 'self-born'. I am tempted to leave that question as an exercise for the reader, but cannot bring myself to be so unkind. The answer is all the way to 10^{16}.

8. NUMBER PATTERNS AND SYMMETRIES

Integers which read the same backwards and forwards are called *palindromes*. Some simple examples are 1,111, 12,321, and 1,397,931. Palindromes make up a rather small fraction of all numbers and they become progressively scarcer as the numbers get bigger. Thus, while every integer between 1 and 9 is necessarily palindromic in a trivial sort of way, only ten percent of the numbers between 10 and 1,000 are palindromes, and only one percent of those between 1,000 and 100,000 have this property. More generally, the probability of finding a palindromic number falls by a factor of ten every time we increase our number region of interest by a factor of 100. Nevertheless it is quite clear that we can easily make up a palindromic number of any length so that they can undoubtedly be found up to the largest numbers imaginable.

We single them out here not only because they have a simple symmetry of digits which makes them attractive to look at, but also because they are involved in a conjecture (as yet unproven and quite possibly false) which I find rather fascinating. It is that if we take any number and add it to the same number with its digits reversed, and then do the same thing with the resulting number and so on, we shall always eventually generate a palindrome. Sometimes the palindrome appears quickly as in

$$27 + 72 = 99,$$

while in others it takes a little longer as with

$$68 + 86 = 154$$
$$154 + 451 = 605$$
$$605 + 506 = 1,111$$

and in still others much longer. Although all integers up to 100 are known to satisfy this speculation, some are stubborn and require a great many steps to generate the palindrome, which is consequently a number of considerable length. The number less than 100 which requires the most steps is 89 which after 24 summations eventually produces the palindrome

$$8,813,200,023,188.$$

This is merely an early indication that, in the absence of any general proof of the palindrome conjecture, direct verification to larger numbers may turn out to be extremely tiresome.

Not all numbers lead to *different* palindromes. For example, 18, 27, 36, 45, 54, 63, 72, 81, and 90, all produce 99 in a single step. They are therefore said to *belong to* 99, and it is customary to write statements of this kind in the form '18(1) belongs to 99', the 1 in parentheses signifying that 99 is obtained in a single step. In this language we see, from the above examples, that 68(3) belongs to 1,111 and that 89(24) belongs to 8,813,200,023,188. The smallest member of a family of numbers which all form the same resultant (not necessarily palindromic) after only one stage of the addition sequence is referred to as the *basic integer* of this set. Only these basic integers need be tested further for palindromic properties. On the other hand, not all basic integers produce different palindromes since other sequences may become identical at later stages of the game.

When proceeding beyond 100, the process of checking out the numbers does become quite time consuming. It proves necessary to test 180 separate numbers (that is basic integers) with 3-digits, 342 with 4-digits, and no less than 3420 with 5-digits. To my knowledge no investigator has yet proceeded in a systematic fashion beyond 100,000. For integers between 100 and 10,000 it is now known that, while most do form palindromes in relatively few steps, some 249 integers (deriving from 15 basic integers) lead eventually to five separate sequences which have not yet generated a palindrome in the first 10,000 addition steps. Rather surprisingly none of the palindromes located so far for the numbers between 100 and 10,000 is found to take more than the 24 steps which were necessary for the 2-digit number 89. Given this situation it begins to look as if the original conjecture may be false, although we cannot yet be certain. Interestingly some very 'near misses' are recorded in each of the five sequences which have avoided a palindromic termination point up to the ten thousandth summation step. For example

$$7059(6) = 46\ 928\ 64$$

by which we mean that the sixth summation step, starting from 7059, produces the nearly palindromic number 4,692,864 and where we have written the latter above in a digit pattern which emphasizes its near palindromic property. In like manner, for the four other sequences, we have

$$879(8) = 88\ 847\ 88$$

$$1997(15) = 935\ 23232\ 638$$
$$196(16) = 897\ 100\ 798$$
$$9999(46) = 3863\ 706\ 276\ 57675\ 672\ 706\ 3683.$$

For numbers greater than 10,000 some palindromes have been found beyond the twenty fourth step, which seemed to be the 'sticking point' for the lower number sequences. The largest and 'most delayed' palindrome known at the time of writing is

$$10911(55) = 4668\ 7315\ 9668\ 4224\ 8669\ 5137\ 8664.$$

Nevertheless, no fewer than 5,842 5-digit numbers (which is between 6 and 7 percent of all the 5-digit numbers) have no palindromes at least in the first 10,000 summation steps. These make up 69 distinct 'non-palindromic' sequences which, together with the 5 sequences discussed earlier for the smaller numbers, give a grand total of 74 such sequences remaining for all integers less than 100,000. Once again, some amusing near palindromes are found in the 5-digit sequences, perhaps the most intriguing being a consecutive pair in the sequence generated by 80,289 namely

$$80289(22) = 27\ 9988\ 6655\ 8899\ 72$$
$$80289(23) = 55\ 9977\ 2222\ 7799\ 44.$$

The percentage of 1-digit, 2-digit, 3-digit, 4-digit, and 5-digit integers producing palindromes in less than ten thousand operations is respectively

1-digit	100%
2-digit	100%
3-digit	98.6%
4-digit	97.4%
5-digit	93.5%

which is a decreasing sequence. Moreover, since the largest palindrome yet produced has 'only' 28-digits, whereas the remaining 74 sequences (for numbers less than 100,000) are free of palindromes out to a few hundred digits, it does seem to be extremely probable that the original hypothesis is false, particularly when we recognize that palindromic numbers become progressively scarcer as we move to larger and larger digit lengths anyway. On the other hand, a few hundred digits is a long way from infinity so perhaps all is not lost and, for the believer, there remains the fact that no pattern has yet been discovered in any of the series which would assure that no palindrome is possible. Interestingly, none of the 74 remaining sequences has shown a tendency to converge with any other over the thousands of addition steps so far studied. One can speculate, of course, whether any will eventually converge and many possibilities remain open. Finally, although the palindromic conjecture remains unproven or disproven for the familiar decimal numbers, it is known to be false in the base-2 counting system.

Before leaving palindromes altogether it is interesting to note that they have also arisen in another context. Thus, while most palindromic numbers do not remain palindromic when squared or cubed, although some do, like

$$111^2 = 12321$$

$$111^3 = 13\ 676\ 31,$$

it is true that most numbers which are palindromic *and* are squares (or cubes) have square roots (or cube roots) which are also palindromes. For squares, although the above statement is true, it is nevertheless not difficult to find some exceptions. The simplest is

$$676 = 26\times26$$

while another is

$$698,896 = 836\times836.$$

For cubes, however, the result is stunning. A computer check of all cubes up to 2×10^{14}, which is two hundred trillion, has found only one palidromic cube which does not have a palindrome for its cube root. It is

$$106\ 62526\ 601 = 2201\times2201\times2201.$$

Given this astounding fact it is tempting to wonder whether 2,201 is the *only* non-palindromic integer which has a palindromic cube. At the moment the answer (like so many others concerning the 'simple' counting numbers) is not known.

Continuing this pattern we can now proceed to look for fourth and higher powers with palindromic properties. Equally astonishing is the fact that a computer check of palindromic fourth powers (again up to two hundred trillion) has failed to turn up a single example which does not have a palindromic fourth root. Moreover, these fourth roots are all of the form 10....01, the smallest being

$$1001^4 = 10040\ 060\ 04001.$$

As for palindromic fifth powers (or higher powers at least up to the tenth) there seem to be none at all as far as the computer search has gone. In fact it has now been proposed that there are no palindromes at all among the numbers of finite size which are fifth or higher powers.

Palindromes, of course, are not the only numbers which have interesting digit patterns. Many of the properties of more common integers, when combined or manipulated by the ordinary rules of arithmetic, can reveal all sorts of remarkable symmetries. As a simple example we might offer the series of equations

$$
\begin{array}{rcccc}
1 \times 8 & + & 1 & = & 9 \\
12 \times 8 & + & 2 & = & 98 \\
123 \times 8 & + & 3 & = & 987 \\
1234 \times 8 & + & 4 & = & 9876 \\
\end{array}
$$

and so on all the way down to

$$123456789 \times 8 \quad + \quad 9 \quad = \quad 987654321$$

which is quite a striking pattern you must admit! There are many others, just two of which are

$$
\begin{array}{rcr}
11^2 & = & 121 \\
111^2 & = & 12321 \\
1111^2 & = & 1234321 \\
\end{array}
$$

and so on to

$$111111111^2 \quad = \quad 12345678987654321$$

and

$$
\begin{array}{rcr}
3 \times 37 & = & 111 \\
6 \times 37 & = & 222 \\
9 \times 37 & = & 333 \\
\end{array}
$$

and so on to

$$
\begin{array}{rcr}
24 \times 37 & = & 888 \\
27 \times 37 & = & 999 \\
\end{array}
$$

but those I like the best are ones which you can invent for yourself;

always different and (if you know the tricks) likely to impress your audience. Some of my favorites are known as *multigrades*. You may not have realized, for example, that

$$10^2 + 11^2 + 12^2 = 13^2 + 14^2$$

or that

$$21^2 + 22^2 + 23^2 + 24^2 = 25^2 + 26^2 + 27^2$$

$$36^2 + 37^2 + 38^2 + 39^2 + 40^2 = 41^2 + 42^2 + 43^2 + 44^2.$$

Some expressions of this general kind (in particular those which contain an equal number of terms on each side) can be formed at will. An easy method is to start off with a simple and obvious equality like

$$3 + 7 = 2 + 8.$$

Proceed by adding any constant number (say 3) to each term to give a second equation

$$6 + 10 = 5 + 11$$

which, of course, is still correct.

A second order multigrade (that is one involving the squares of numbers) can now be formed from the two equations above by 'switching sides' and combining as follows:

$$3^2 + 7^2 + 5^2 + 11^2 = 6^2 + 10^2 + 2^2 + 8^2.$$

Checking to see whether the 'switching sides' recipe really works, we find that the left hand side is

$$9 + 49 + 25 + 121 = 204$$

while the right hand side is

$$36 + 100 + 4 + 64 = 204$$

so that all is well. You can verify that there is nothing special about these particular numbers by choosing others and doing it again. In fact, any reader who is familiar with simple algebra can easily check for himself that the method works in the general case. This being so we can make the result look a little more impressive (and 'magic') by choosing numbers so that the equation reads the same from either end (that is to say is palindromic). For example, start with

$$45 + 32 = 23 + 54$$

which is both true and palindromic. By adding 11 to each term on both sides of the equation, we obtain a second true statement

$$56 + 43 = 34 + 65$$

which is also a palindrome. We are now all set to use the 'switching sides' trick once more to get

$$45^2 + 32^2 + 34^2 + 65^2 = 56^2 + 43^2 + 23^2 + 54^2.$$

This is a *palindromic multigrade* to give it its proper name, and is quite true, each side adding up to 8,430. With a little practice you will find it easy to make up other palindromic multigrades of this kind. Still more variations of this game are possible. For example, we can restrict the multigrade to prime numbers. But the really nice thing about the crossing-over routine is that it can be carried on to multigrades involving higher powers; cubes, fourth powers, and so on.

Adding any integer to each number which is squared in a second order multigrade produces, amazingly, another second order multigrade which is still true. Let us take the palindromic multigrade set out above and add 11 to each number which is squared, in order to maintain the palindromic property. We obtain the new form

$$56^2 + 43^2 + 45^2 + 76^2 = 67^2 + 54^2 + 34^2 + 65^2$$

in which each side now adds up to 12,786. The crossover routine can now be applied to these two palindromic multigrades to produce a palindromic third order (or cubic) multigrade. It reads

$$45^3 + 32^3 + 34^3 + 65^3 + 67^3 + 54^3 + 34^3 + 65^3$$

$$= 56^3 + 43^3 + 45^3 + 76^3 + 56^3 + 43^3 + 23^3 + 54^3.$$

Any readers who doubt this result can extract their calculators and verify the truth of this palindromic wonder; each side adds up to 1,209,978. In addition, by the way it was formed, this equation is still valid if the power 3 is reduced to 2 or to 1, so that we really possess three intimately related palindromic forms, one linear, one quadratic, and one cubic. These equations certainly do read the same from both ends, but the perfectionist may object to the fact that all the numbers in it are not different. In particular 45^3 and 54^3 can be subtracted from both sides of the cubic multigrade to produce

$$32^3 + 34^3 + 65^3 + 67^3 + 34^3 + 65^3$$

$$= 56^3 + 43^3 + 76^3 + 56^3 + 43^3 + 23^3.$$

This is really an advantage, since it means that we are not restricted to generating third order multigrades which always have eight terms on each side. But there are still two 34's and two 65's on the left hand side paired with two 43's and two 56's on the right hand side. There is no fundamental difficulty in rectifying this remaining 'aesthetic fault' and if the reader wants to go back and start the process again with a more thoughtfully chosen set of numbers he will doubtless be able to succeed.

As mentioned before, the above procedure can be continued ad infinitum, and at each stage the multigrade is valid not only for the power reached at that stage, but for all the lower powers as well. Knowing the crossover construction technique makes the formation of countless numbers of these rather impressive forms quite routine.

Finally I shall close this chapter with a few other striking patterns in number properties which should not be allowed to pass by without mention. They involve the so-called *narcissistic* numbers (that is, numbers which are 'wrapped-up' in themselves). This concept is made clear by the following examples:

$$153 = 1^3 + 5^3 + 3^3$$

$$371 = 3^3 + 7^3 + 1^3$$

$$407 = 4^3 + 0^3 + 7^3.$$

A somewhat more impressive one is

$$165{,}033 = 16^3 + 50^3 + 33^3$$

and, in a variation on this same theme, we also find

$$81 = (8 + 1)^2$$

and

$$4{,}913 = (4 + 9 + 1 + 3)^3.$$

9. NUMBERS PERFECT, FRIENDLY, AND WEIRD

Ever since Greek times the so-called perfect numbers have been objects of numerological speculation. In 2000 years no great use has ever been found for them, but they continue to mystify and fascinate people who are interested in the 'personality' of integers. Every integer seems to have a distinct character, but the label of *perfection* is not to be bestowed without due consideration, and the origin possibly goes back to biblical times and the statement that God created the world in six days. Six, you see, is the first (that is smallest) perfect number, and this perfection surrounds the fact that six is equal to the sum of its *divisors,* which are all those integers which divide it exactly (namely 1, 2, and 3). What is more, for this case alone the product of these divisors (that is to say 1×2×3) is also equal to the same number 6 and, as Isaac Asimov has said, 'God could hardly be expected to resist all that'.

Most integers are said to be *deficient* numbers in the sense that their divisors add up to less than the number itself. Counting up from 1 we find, with the exception of the perfect 6, that all the integers are deficient (the primes very badly so) until we get to 12. The number 12, on the other hand, with divisors 1, 2, 3, 4, and 6, which add up to 16, is certainly not deficient in this sense. With divisors which add up to more than the parent integer, it is called an *abundant* number. The next largest abundant numbers are

$$18, 20, 24, 30, 36, ...$$

so that these, although not as common as deficient numbers, are hardly rarities. In fact there are 21 abundant numbers up to 100 and they are all even. Does this mean that all abundant numbers are even, you may ask? The answer is nearly, but not quite all. The smallest abundant odd number is 945 and has divisors

$$1, 3, 5, 7, 9, 15, 21, 27, 35, 45, 63, 105, 135, 189, 315$$

which add up to 975. Surprisingly, although odd abundant numbers are comparatively rare, there is no difficulty whatsoever in locating arbitrarily large ones. This follows from the fact that *any* multiple of an abundant number is also abundant. This means that 945, for example, can be multiplied by any odd number whatsoever, and the result is also an abundant odd number. For this reason the abundant odd numbers carry little mystique.

The perfect numbers, on the other hand, have considerable mystique since they are rarer by far than any classes of deficient or abundant numbers. The first few are

$$6; \quad 28; \quad 496; \quad 8,128; \quad 33,550,336; \quad ...$$

and the gaps between them seem to be widening rapidly. The first obvious question is 'are there infinitely many of them or is there a largest one of all?' At present the answer to this question is not known but, surprisingly, one general type of perfect number has been classified since Greek times; by Euclid none the less. In the ninth book of Euclid's *Elements* one finds the proof that any number of the form

$$2^{p-1} \times (2^p - 1)$$

is perfect if both p and $2^p - 1$ are prime numbers. It does not say that all perfect numbers have to be of this form, but all those known at the time of writing are. Thus, for example,

$$6 = 2 \times (2^2 - 1)$$

$$28 = 2^2 \times (2^3 - 1)$$

$$496 = 2^4 \times (2^5 - 1)$$

$$8,128 = 2^6 \times (2^7 - 1)$$

$$33,550,336 = 2^{12} \times (2^{13} - 1).$$

But wait a moment! Haven't we seen the prime numbers $2^p - 1$ before? Indeed we have; they are the Mersenne primes which led the drive for finding the world's largest prime number. The largest known in 1980, as recorded in Table 2 of Chapter 3, is for $p = 44,497$. It therefore generates a perfect number of truly gigantic proportions, namely

$$2^{44496} \times (2^{44497} - 1)$$

with no fewer than 26,790 digits. Since the Mersenne prime $2^{44497} - 1$ is only the twenty seventh to be discovered, this largest known perfect number is only the twenty seventh perfect number yet discovered. Therefore we can easily tabulate all the known perfect numbers and this is done in Table 6, where we give the relevant prime p and the length (in digits) of the perfect number which is generated, via Euclid's formula, using this prime.

TABLE 6		
The 27 Known Perfect Numbers $2^{p-1}\times(2^p-1)$		
Generating Prime p	Number of Digits	Year of Discovery
1 — 2	1	ancient (B.C.)
2 — 3	2	ancient (B.C.)
3 — 5	3	ancient (B.C.)
4 — 7	4	before A.D. 100
5 — 13	8	circa 1456
6 — 17	10	circa 1588
7 — 19	12	circa 1588
8 — 31	19	circa 1750
9 — 61	37	1883
10 — 89	54	1911
11 — 107	65	1914
12 — 127	77	1876
13 — 521	314	1952
14 — 607	366	1952
15 — 1279	770	1952
16 — 2203	1327	1952
17 — 2281	1373	1952
18 — 3217	1937	1957
19 — 4253	2561	1961
20 — 4423	2663	1961
21 — 9689	5834	1963
22 — 9941	5985	1963
23 — 11213	6751	1963
24 — 19937	12003	1971
25 — 21701	13066	1978
26 — 23209	13973	1979
27 — 44497	26790	1979

As mentioned before, all the known perfect numbers are of Euclid's form. We might ask how much of a coincidence this is since it is not known that all perfect numbers have to be of this form. Well, in spite of the enormous efforts which have been concentrated on the Mersenne primes in the race for the largest known prime number, it is not wholly coincidence since Euler, some two thousand years after Euclid, made the second great breakthrough in perfect number theory. He succeeded in proving that all *even* perfect numbers *must* be of the Euclidean form. Whether odd perfect numbers exist is still not known, but quite evidently if they do they must be very rare. Just how rare can be judged from the fact that an exhaustive search to 10^{200} has not unearthed a single one, so that they are truly of the utmost rarity. But even 10^{200} is a long way from infinity so that a fascination concerning the possibility of discovering the first odd perfect number remains. Some results are known concerning odd perfect numbers if they do exist. For example, it is known that they must be divisible by at least 7 distinct primes, the largest of which must be greater than 100,000 and the second largest of which must be greater than 139. On the other hand much more is known about the even perfect numbers, which unquestionably do exist.

Firstly, all even perfect numbers can be derived as a sum of the form

$$1 + 2 + 3 + 4 + 5 + \ldots$$

which mathematicians call an *arithmetic progression*. For example

$$6 = 1 + 2 + 3$$
$$28 = 1 + 2 + \ldots + 6 + 7$$
$$496 = 1 + 2 + \ldots + 30 + 31$$
$$8,128 = 1 + 2 + \ldots + 126 + 127$$

where you will notice that the last term in each progression is the Mersenne prime $2^p - 1$ from which the perfect number is obtained via Euclid's formula. Secondly, an even more remarkable finding concerns the *reciprocals* of the divisors of a perfect number, where the reciprocal of any number is defined as 1 divided by that number. It is that the sum of the reciprocals of all the divisors of the perfect number (this time and only this time including the perfect number itself) is always exactly equal to 2. For example, since the divisors of 6 in this context are 1, 2, 3, and 6, we observe that

$$(1/1) + (1/2) + (1/3) + (1/6) = 2.$$

Similarly, for the perfect number 28, with divisors 1, 2, 4, 7, 14, and 28, we find

$$(1/1) + (1/2) + (1/4) + (1/7) + (1/14) + (1/28) = 2$$

and so on for the others. Thirdly, all the perfect numbers end in a 6 or an 8. There is, however, no obvious pattern in the manner in which the 6's and the 8's occur for increasingly large perfect numbers. The pattern so far for the 27 known 'perfects' is

686 866 886 688 688 866 686 666 666

with the sixes leading by 17 to 10 and seemingly pulling away. Will the eights ever catch up? I am sure that your guess is quite as good as that of any mathematician alive.

With so few perfect numbers to play with, the concept of an 'almost perfect number' has also been introduced. An almost perfect number is one for which the divisors add up to just one less or one more than the number in question (now no longer including the number itself in its divisors). Let us first concentrate on the almost perfect but slightly deficient numbers. One complete class of such numbers is easy to establish; it is all the powers of 2. For example

$$2^2 = 4 \text{ has divisors } 1,2 \text{ which add up to } 3$$

$$2^3 = 8 \text{ has divisors } 1, 2, 4 \text{ which add up to } 7$$

$$2^4 = 16 \text{ has divisors } 1, 2, 4, 8 \text{ which add up to } 15$$

and so on. But it is far from simple to find any other examples which are not of the form 2^n. In fact at the time of writing, with the exception of these powers of 2, no almost perfect numbers are known which are either deficient or abundant. On the other hand it is known that any almost perfect abundant number (if it exists) must be the exact square of an odd number. It must also be at least as large as 10^{20}.

Another category of 'perfect' numbers, recognized as long ago as 1631 by Mersenne, is the so-called class of *multiply perfect numbers*. For these the sum of the divisors of the number is exactly 2, 3, 4, or more times the number itself. Correspondingly we refer to such numbers as being perfect numbers of class 2, 3, 4, ... etc., where the conventional perfect numbers are, in the extended notation, now perfect numbers of class 1. The smallest multiply perfect number is 120 with divisors

1, 2, 3, 4, 5, 6, 8, 10, 12, 15, 20, 24, 30, 40, 60

which add up to 240, or two times 120. The number 120 is therefore the smallest perfect number of class 2. The next smallest multiply perfect number is also of class 2; it is 672. Many others are now known, in fact hundreds of them, with some as high as class 7. The smallest perfect number of class 3 is

$$30,240 = 2^5 \times 3^3 \times 5 \times 7$$

and the smallest of class 4 is

$$14,182,439,040 = 2^7 \times 3^4 \times 5 \times 7 \times 11^2 \times 17 \times 19.$$

Higher classes are made up of still larger numbers and we might ask whether multiply-perfect numbers exist up to arbitrarily high classes. It seems possible but I know of no proof of this fact.

AMICABLE NUMBERS

We have seen that all integers can be classified as deficient, perfect, or abundant according to whether the sum of their divisors is smaller than, equal to, or larger than the number itself. Among all numbers classified in this way by the ancient Greeks one very special pair was noticed. It is the pair

$$220, \quad 284$$

of which the former is an abundant number and the latter a deficient one. If we write down and add the divisors of each, we can readily discern for ourselves what it was that the Greeks found to be so special about them. Thus, 220 is exactly divisible by

$$1, 2, 4, 5, 10, 11, 20, 22, 44, 55, 110$$

and these numbers add up to 284. On the other hand 284 has the divisors

$$1, 2, 4, 71, 142$$

which add up to 220. It follows that the sum of the divisors of each is exactly equal to the other.

Is such a situation rare, you ask? Well it was certainly sufficiently rare that the Greeks knew of only this single example. They termed the pair *amicable,* or friendly, numbers since (as Pythagoras, who is credited with their discovery, affirmed) each has the power to generate the other according to the rules of friendship. No other pair of amicable numbers was found until the year 1636 (by which time seven perfect numbers were already known) when the redoubtable Fermat announced a second pair

$$17,296, \quad 18,416.$$

This pair was not discovered by chance, but followed from a specific algebraic formula which had first been given by Arabian mathematicians in the ninth century, and there is some evidence that Arab mathematicians may themselves have known of this second amicable pair as early as the thirteenth century (although Fermat is still usually credited with the discovery in the textbooks). This Arabian formula only generated actual amicable pairs if certain types of prime number could be located. In this sense it was like the Euclidean form for perfect numbers. A continued pursuit of the same formula also provided the third known amicable pair, which was announced in the year 1638 by the French mathematician Rene Descartes. This pair was

$$9,363,584, \quad 9,437,056.$$

The sheer size of this third pair may have suggested that the amicables were of the greatest rarity, perhaps even rarer than the perfect numbers. Such has not turned out to be the case, in spite of the difficulty that the earlier researchers experienced in trying to find them. The formula pursued by Fermat and Descartes did not promise to give *all* the amicable numbers, and it is now known that there are several other examples smaller than Fermat's pair and several hundred smaller than Descartes' pair. Thus, unlike the case with the perfect numbers, other formulas do exist which can generate amicables. The first real breakthrough in locating amicable pairs came from the meticulous Euler who, with customary thoroughness, proceeded to discover no less than 59 pairs between the years 1747 and 1750. After a comparative drought lasting for over two millennia this was indeed a flood. The smallest pairs were

$$2620 = 2^2 \times 5 \times 131; \quad 2924 = 2^2 \times 17 \times 43;$$
and
$$5020 = 2^2 \times 5 \times 251; \quad 5564 = 2^2 \times 13 \times 107.$$

Euler did such an excellent job that only four more examples were discovered in the next one hundred and fifty years, bringing the total to 66 by the end of the nineteenth century. But among these four additional pairs is one which deserves a special mention. It illustrates the romance that computational mathematics contained before the computer era. In the year 1867, a sixteen-year-old boy named Nicolo Paganini announced (without any indication of the method of discovery) a new amicable pair which was smaller than any discovered before, save only for the original pair 220, 284 which had been known for 2,500 years. This pair

$$1,184 = 2^5 \times 37, \quad 1,210 = 2 \times 5 \times 11^2$$

had somehow escaped detection by the great Euler, and one is tempted to smile at the seemingly absurd optimism of a young lad doodling with number divisors, just on the off chance that in there somewhere might be an amicable duo which all the great mathematicians before him had missed. It is now known that Paganini's amicable pair is the smallest which exists other than the original 220, 284.

With the aid of modern-day computers all the amicable pairs up to 100 million have now been located. No longer can the enthusiastic amateur search for any which might have been forgotten. It turns out that Euler had found all the remaining pairs smaller than 100,000 with only three exceptions. There are, including the original 220, 284 and Fermat's pair, thirteen such friendly pairs, and the three which were unknown to Euler were discovered separately at wide intervals. First came the stunning find of the youthful Paganini in 1867. Some 72 years later in 1939 the penultimate pair was published by B. H. Brown at a time when no less than 389 other pairs were already known, some as large as 10^{24}. Yet, incredible as it may seem, one pair smaller than 10^5 had still eluded detection, and did so right up to the time when a full computer search was finally made in 1966 to locate all amicables up to one million. This final reluctant pair was

$$79,750 = 2 \times 5^3 \times 11 \times 29, \quad 88,730 = 2 \times 5 \times 19 \times 467$$

and was one of 9 pairs smaller than a million which were unearthed for the first time in the final computer assault. We list the 15 smallest amicable pairs, together with their discoverers, in Table 7.

Since 1966 the computer search for amicable pairs has been continued and it is now known that there are 42 pairs with the smaller number less than 1 million, 108 pairs with the smaller number less than

TABLE 7			
The smallest pairs of Amicable Numbers N and M, together with their discoverers and the respective ratios M/N (to 2 significant figures).			
N	M	Discoverer	M/N
220	284	Pythagoras (540 B.C.)	1.29
1,184	1,210	Paganini (1867)	1.02
2,620	2,924	Euler (1747)	1.12
5,020	5,564	Euler (1747)	1.11
6,232	6,368	Euler (1747)	1.02
10,744	10,856	Euler (1747)	1.01
12,285	14,595	Brown (1939)	1.19
17,296	18,416	Fermat (1636)	1.06
63,020	76,084	Euler (1747)	1.21
66,928	66,992	Euler (1747)	1.00
67,095	71,145	Euler (1747)	1.06
69,615	87,633	Euler (1747)	1.26
79,750	88,730	Alanen et al. (1966)	1.11
100,485	124,155	Euler (1747)	1.24
122,265	139,815	Euler (1747)	1.14

10 million, and 236 pairs with the smaller number less than 100 million. Given this information it is now possible to study the manner in which the amicable pairs distribute themselves on average as we go to higher and higher values. In doing this we must recognize that no proof has yet been given that the number of amicables is infinite, so the possibility of there being a largest pair of all cannot be entirely dismissed. However, if there are infinitely many, it does appear from the data up to 10^8 that the number of amicable pairs less than some large number n (let us call this number of pairs $A(n)$) goes something like

$$A(n) = \sqrt{n}/(2ln(n)).$$

On the other hand, if we do not know for certain that the amicables go on forever, it is obvious that we cannot take this formula too seriously beyond the range of the computer data.

There are many interesting speculations concerning these 'friendly' numbers, all substantiated by computer evidence, but none yet established beyond question. The first is that amicables appear either as two even numbers or as two odd numbers; never as one even and one odd. A second is that all the odd amicable numbers have a divisor 3; a third that all amicable pairs contain a common divisor. This last conjecture would follow, of course, if the first two were known to be correct (all even-even amicables being divisible by 2 and all odd-odd amicables by 3). It has been established that if an amicable pair N, M does exist with N even and M odd (or vice versa) then they must be bigger than 10^{37}. It has also been established that if there are odd-odd pairs without a common divisor then they must be larger than 10^{36}. A final conjecture concerning amicables is that as they get larger and larger (if they are in fact boundless) the ratio M/N between the components of a pair gets closer and closer to 1. It can be seen from Table 7 that, even for the smaller pairs, this ratio never gets much larger than 5/4.

The search for the largest known amicable pair has been pursued relentlessly over the years but, for some reason, without the publicity and glamour surrounding similar searches for the largest known prime and perfect numbers. The latter two tend to go hand in hand, as we have seen, since all the even perfect numbers (which are the only ones yet known) are directly related to the Mersenne primes $2^p - 1$ which have consistently led the prime number race over the years. In the amicable business the rules followed are essentially adaptations of one given way back in the ninth century by a certain Arabian mathematician Thabit ibn Kurrah. The original rule stated that two numbers of the form

$$N = 2^n \times p \times q \quad \text{and} \quad M = 2^n \times r$$

are amicable if the three numbers p, q, and r are all prime numbers of the rather special form

$$p = (3 \times 2^{n-1}) - 1,$$

$$q = (3 \times 2^n) - 1,$$

$$r = (9 \times 2^{2n-1}) - 1.$$

The original Pythagorean pair $N = 220$ and $M = 284$, in particular, satisfy this relationship if $n = 2$, leading to values $p = 5$, $q = 11$, and $r = 71$.

If prime numbers of the above form can be found for a large integer n then new and extremely large amicable pairs follow immediately. For smaller values of n some success has been obtained via this method, and both $n = 4$ and $n = 7$ provide 'good' solutions. Fermat used the

$n = 4$ solution to find the first new amicable pair of the modern age (although as we have mentioned it does seem likely that the Arabs themselves knew this solution much earlier). With $n = 4$ we find, from the above equations, the values $p = 23$, $q = 47$, and $r = 1,151$ (all primes as required) generating the amicable pair

$$N = 2^4 \times 23 \times 47 = 17,296$$

and

$$M = 2^4 \times 1,151 = 18,416$$

which can be found in Table 7 as the eighth smallest amicable pair. The $n = 7$ solution was that found by Descartes in 1638 and leads to

$$N = 2^7 \times 191 \times 383 = 9,363,584$$

and

$$M = 2^7 \times 73,727 = 9,437,056.$$

The size of this type of amicable pair evidently increases extremely rapidly with n. The potential for generating large amicable pairs therefore seems to be great except for one unfortunate fact. It is that not a single further value of n satisfying the equations has yet been discovered, although a computer search has been carried out all the way up to $n = 20,000$. Nevertheless all is not lost, because it has been found possible to extend Thabit ibn Kurrah's rule in several different ways which have led to new and extremely large amicable pairs. All these extended rules focus on the ability to locate prime numbers of a special form. Thus the search for amicables, using these extended 'Thabit' rules, always eventually reduces to a test of primeness for certain generated large numbers.

From 1946 to 1974 the record appears to have been held by a pair of 25-digit numbers. In 1974 new amicable pairs of 32, 40, 81, and finally 152-digits were announced resulting from a computer search using the modified Thabit rules. In the 40-digit numbers the first 12 digits of each were identical, leading to a ratio M/N which was extremely close to 1. This seemed to be in accord with the expectation that this ratio should get closer and closer to 1 (at least on the average) as the size of the amicables increased. Unfortunately the 81-digit and 152-digit pairs did not conform. The 81-digit amicable pair had a ratio M/N of 1.030 while the 151-digit monsters had a ratio even further removed from 1 at 1.046. This last conjecture therefore seems to be on somewhat shaky ground at the present time.

One final peculiarity associated with the amicable numbers is that very nearly all the amicables are divisible by 9 when the pair is added. Thus, for example,

$$220 + 284 = 504 = 9 \times 56.$$

As can be seen from Table 7, of the smallest 15 amicable pairs only one (namely 12,285 and 14,595) fails to satisfy this 'divisible by 9' property. But for *even* amicables the rule is so good that it is difficult to find exceptions to it at all. The smallest pair known to me which breaks the rule is

$$666,030,256 \text{ and } 696,630,544.$$

How a rule can be that good without being perfect is quite astounding.

SOCIABLES AND WEIRDS

Recalling the definition of amicable numbers, that the divisors of each should add up to the other, it is a rather natural extension of this idea of 'friendliness' to ask whether larger groups of numbers can be found which are related in a similar fashion. As a particular example we might ask whether there are groups of three numbers, let us label them L, M, and N, such that the divisors of L add up to M, the divisors of M add up to N, and the divisors of N add up to L. Evidently no pair of such a group can be amicable as defined earlier, but it seems only fair to admit that if they exist they should be recognized as a rather sociable trio. Groups of numbers with this chainlike property of divisor association, whether there are 3, 4, 5, ... or even more members in the group, are actually called *sociable* groups by the number theorists. For the particular case of 'sociable groups of three' the word *crowds* is sometimes used.

Most of the known sociable groups seem to contain four members each. At the time of writing no three number groups or 'crowds' have been located although no-one has proved that they do not exist. Among the sociable groups of 'order-4' (that is having four members) the one involving the smallest numbers seems to be

$$1,264,460; \quad 1,547,860; \quad 1,727,636; \quad 1,305,184.$$

A computer search for sociable groups with 10 or fewer members has been made for all numbers up to 60 million. Ten groups were located in this search and, very surprisingly, all except one are of order-4. The odd one out is of order-5 and contains relatively small numbers. For the record it is

$$12,496; \ 14,288; \ 15,472; \ 14,536; \ 14,264.$$

This group had in fact been known before the computer search was made, and other groups with more than 10 members do exist, the largest being one with no less than 28 members. On the other hand the interest in sociables is of relatively recent origin compared with the perfect and amicable number research. As a result the number of sociable groups so far discovered is far less than the known number of amicable pairs. There is every probability that much more will be learned of these particular groups when more computer effort is brought to bear upon the problem in the future. In the meantime the most interesting question seems to be 'are there any crowds out there?'

Enough! I hear you cry; surely we have now exhausted the discussion of these weird numbers. But no; it turns out that we have not yet touched upon the truly weird numbers. You see there are such objects, actually named *weird* numbers by the mathematicians, and I cannot let this chapter pass by without making at least some passing reference to them (for your pleasure). You will recall that those numbers like 12, which have divisors adding up to more than the number itself, are called 'abundant' numbers. Now it happens that nearly all (but not quite all) the abundant numbers have among their divisors a smaller subset which adds up exactly to the parent number. This does not make them perfect, of course, because there are other divisors left over, but we can demonstrate the situation by considering a couple of small abundant numbers like 12 and 30.

Now the integer 12 has divisors 1, 2, 3, 4, and 6, which add up to 16. Since 16 is larger than 12 we verify that 12 is abundant. Moreover we can choose from the total number of divisors a smaller selection, namely 2, 4, and 6 (or equivalently 1, 2, 3, and 6) which do add up to 12. In like manner the integer 30 has divisors 1, 2, 3, 5, 6, 10, and 15, which add up to 42. Again we can find a combination from this set (for example 5, 10 and 15) which exactly adds up to 30.

All this is regarded by number enthusiasts as being normal and decidedly unweird. However, one can now ask whether any abundant numbers exist which do not have this property. As stated above there are such numbers, but they make up only a small fraction of all the abundant numbers. Perhaps because of this relative scarcity (or possibly because of the rather limited vocabulary of some mathematicians) they are termed 'weird' numbers, and there are only two smaller than 1000. The smallest of all is 70, which has divisors

$$1, 2, 5, 7, 10, 14, 35.$$

These add up to 74 so that the number 70 is indeed abundant. However, there is no combination of divisors among the group set out above which exactly adds up to 70. It is therefore 'weird'. The only other weird number less than one thousand is 836. Its divisors are

$$1, 2, 4, 11, 19, 22, 38, 44, 76, 209, 418$$

and add up to 844. It is now easy to verify the weirdness of 836 for yourselves.

At first sight it would seem that abundant numbers are most likely to be weird if they are only just abundant; that is when the sum of all the divisors is only just a little bit larger than the number itself. Whether there is a critical ratio, equal to the sum of all the divisors divided by the number, above which weird numbers cannot occur is not known. Another peculiar property of weirds is that all the known weirds are even. Whether this is an essential condition for weirdness is also not known, but it does seem that odd weird numbers (if they exist) must be of the greatest rarity.

Although the weird numbers are a very small fraction of all the abundant numbers in any range of interest, it is nevertheless possible to establish that they do exist to arbitrarily high values. In other words we do know that there is no largest weird number of all; they do go on for ever like the prime numbers. This is known because it can be established that if a number n is weird, and p is any prime number which is larger than n plus the sum of its divisors, then n times p is also weird. Thus, for example, the weird number 70 times any prime number larger than 144 is also a weird number. This immediately gives us an infinite set of weirds. Notice, however, that they are all even. *Excluding* this series, built up from the smallest weird number 70, there are only 13 weirds which are smaller than 100 thousand; they are

70	836	4,030	5,830
7,192	7,912	9,272	10,792
17,272	45,356	73,616	83,312

and finally

91,388.

10. HOW DO THESE SERIES END?

In the earlier chapters we have already encountered many sequences of numbers which may or may not go on forever. Particularly famous examples are the perfect numbers and the amicable numbers, and others, somewhat less well known, are the emirps and the prime twins. Obviously these sequences are in actuality either finite or they are not; the uncertainty simply reflects our present lack of knowledge concerning them. In some rather embarrassing cases, such as the odd perfect numbers, odd weirds, and crowds, we do not even known whether the sets possess any members at all, although they could quite easily contain an infinite number. Such is the degree of our ignorance. In this chapter we focus on these kinds of difficulties by setting up even more simply defined sequences of integers and asking once again whether these series go on forever, or whether at some point they stop. In some, as is the case with the perfect and amicable numbers, the answers remain unknown, while in others some insight has been achieved.

What is fascinating about many of these series is that they are so easy to set up and yet so difficult to master. A good place to begin is with a very elementary series in which the rule for generating terms is extremely simple in principle. The idea is to start with any integer whatsoever and to generate a series in which each term after the first is the sum of the divisors of the preceding term. Consider, as an example, the series which starts with the integer 12. Now 12 is exactly divisible by

$$1, 2, 3, 4, 6$$

and these divisors add up to 16. The second term in the series is therefore 16. We now proceed to find the divisors of 16, which are

$$1, 2, 4, 8$$

and add these to get 15. The third term in the series is therefore 15 and, with the procedure for continuing the series now well understood, we can find the divisors 15 and continue on our merry way. What happens? Well, a bit of further study reveals that this particular series quickly comes to an end in the following manner:

$$12, 16, 15, 9, 4, 3, 1.$$

Obviously once we obtain a 1 in the series the interest has all gone because the number 1 has no divisors (since we always exclude the number itself from its divisors) and we can progress no further. Checking similar series starting with other small integers it soon becomes apparent that, while some of these series do increase at first, most go straight down and all seem finally to end with a 1 after a comparatively short time. The shortest series of all are obtained for the prime numbers p. These contain only the two terms p and 1. The question we wish to ask is whether *all* such series, starting with *any* integer whatsoever, eventually end with a 1. Surprisingly, the answer to this question is easy to establish because one special set of starting numbers, namely the perfect numbers, has a unique (if rather uninteresting) behavior which does not terminate with a 1. To see what this behavior is we can start with the perfect number 6 and obtain the sequence

$$6, 6, 6, 6, 6, 6, 6, \ldots..$$

and so on forever. It is clear that all perfect starting numbers will repeat themselves in a similar fashion. But this is not very exciting and obviously contains no challenge at all to our intellect. We therefore exclude the perfect numbers from our starting integers (after all there are only 27 of them known to date) and ask the question again. But still the answer is no since any amicable (or sociable) starting number will obviously cycle forever with a period of two (or more), the simplest example being

$$220, 284, 220, 284, 220, 284, \ldots$$

We must therefore exclude these as starting numbers as well and, having done so, pose the question one final time. Excluding sequences which begin with perfect, amicable, or sociable numbers, do all the series made up in the manner set out above finish, after a finite number of terms, with a 1? This time the answer is much more difficult to obtain. What are the possibilities? The series could go on and on with terms increasing forever, or perhaps the divisors somewhere down the line may add up to a perfect (amicable or sociable) number which will then repeat itself (or cycle) forever. To date, however, no examples of cycling or of the appearance of a perfect number inside such a series has ever been discovered.

From the earliest efforts to wrestle with this particular problem there was a general feeling among mathematicians that all such series probably do indeed end with a 1 and that, consequently, none would go on forever, cycling or not. The tests for starting integers up to and including 137 were all quickly carried out and terminating 1's were found. Some smaller numbers do have an initial tendency for generating a series of increasing numbers, for example 30 starts off with

$$30, 42, 54, 66, 78, 90, 144, ...$$

but they all eventually seem to turn over and fall back to finish with a 1. The first number historically to cause any real trouble in the testing program was 138. It proceeds to generate no less than 117 terms before reaching a maximum at

$$179,931,895,322.$$

Computations of this kind, as you can well imagine, required an unusual amount of patience and time in the years before the computer. Eventually, however, the 138 series does begin to fall, and finally it conforms with all its predecessors by terminating with a 1 at the 177th generated term.

The next really bothersome starting number is 276. Its series increases all the way up to the 172nd generated number (which has no less than 28 digits) before turning over. But it also creates another precedent. This time, after passing its peak, instead of producing continuously decreasing terms all the way down to 1, it decreases for a while (actually to the 226th term) but then begins to increase again. We therefore learn that these series can be more complex in form than first suspected. At the 226th term the local minimum in the series still has 15 digits, and beyond this point a rather erratic upward trend is established which has been followed by computer calculation for a further 200 terms, by which time it has reached a term-size involving 36 digits. At the time of writing the known terms are still increasing and, as a result, there is no evidence yet that the series starting with 276 is going to end very soon. Any assertion that all such series finally end in a 1 must therefore be considered to be debatable at the very least.

More research with larger starting numbers has established that there are no less than 6 distinct series involving 13 starting numbers (some series joining together with others as they progress) which begin with integers less than 1000 and reach generated values larger than 30 digits with, needless to say, no sign of any imminent termination. There are also some 98 distinct series (involving 751 starting integers) which start with numbers less than 10,000 and reach the 24-digit level. It is now thought quite possible that many of these may be 'unbounded', that is go on forever upward without repeating or cycling. But such remains to be established and is today just one more unproven assertion involving the counting numbers.

As a second example of a series of numbers which can be set up using a simple rule, and which can be a bit deceiving, I ask you to consider the following. Note first that any single digit number is (trivially) divisible by 1; there are therefore 10 of them. We now ask how many 2-digit numbers are divisible by 2, and a quick check reveals that there are 45 of them (ignoring initial zeroes). The next step of the process requires the computation of the number of ways in which these 45 2-digit numbers can be extended to 3-digit numbers which are exactly divisible by 3. Again a little computation delivers the answer; it is 150.

Suppose that we generalize this procedure and ask how many numbers there are in which the first n digits are exactly divisible by n, with n taking values from 1 all the way up to the length of the number. Evidently, from the consideration already given to the problem, the series begins

$$10, \quad 45, \quad 150, \quad ....$$

and the continuation is just a matter of 'simple' computation. We now wish to ask the question 'does this series go on forever or does there come a time when it ends with a zero, implying that at some digit length no integer at all satisfies all the divisibility criteria?' It is easy to verify that the series must continue on for several more terms at least, since at any point one can always find a suitable extra digit to add on, at least up to a digit length of ten. For example, suppose that we start with 2. For the second digit we can choose any even numeral (to ensure divisibility by 2). Suppose we pick a 4 to make 24. We can now add a zero (or a 3, or 6, or 9) to produce a 3-digit number divisible by 3, and so on. The point is that since we have a choice of ten numerals 0 to 9 at each step, one of them at least will ensure an exact divisibility by any integer up to ten. We can therefore *always* continue a particular number sequence up to ten digits. For example the sequence starting

2, 24, 240 can be continued as follows:

> 2,   24,   240,   2404,   24045,   240456,   2404563,
> 24045632,   240456321,   2404563210,

verifying that no difficulties arise up to this point. But beyond the tenth digit it is no longer certain that any particular sequence of numbers, such as that set out above, can necessarily be continued. With this particular sequence, for example, it turns out that an eleventh digit can be added to produce a number exactly divisible by 11 (it is 24,045,632,105) but a twelfth digit cannot be found. That is to say none of the numbers 240,456,321,05X, in which X is any numeral 0 to 9, is divisible by 12. We might therefore suspect that the number of N-digit integers in which the first n digits (with n running from 1 to N) are exactly divisible by n could possibly start to decrease when N goes beyond 10.

To check this out properly calls for a computer search. It is found that our series which started for 1-digit, 2-digit, and 3-digit numbers with 10, 45, and 150 respectively, actually reaches a maximum at the ninth and tenth terms which are both equal to 2,492. This means that there are 2,492 different 10-digit numbers like the tenth one in the 2, 24, 240, ... sequence above. As the series continues the terms now begin to fall in magnitude and, by the time we reach the 20th term, it is only 44. In other words, there are only 44 20-digit numbers in which the first n digits are exactly divisible by n all the way out to $n = 20$. For $n = 20$ through $n = 25$ the terms in the series continue to fall in the manner

> 44, 18, 12, 6, 3, 1.

Therefore a unique 25-digit number exists for which the first n digits (with n going all the way from 1 to 25) are exactly divisible by n; it is

> 3,608,528,850,368,400,786,036,725.

Because no 26th digit can be added to this to make a number divisible by 26 the sequence ends at the 26th term with a zero. This series, at least, we now know does not go on forever but, before leaving it, there is one rather unlikely looking member of the group of 2,492 10-digit numbers which is worth recording; it is

> 3,000,060,000.

This is not the kind of digit pattern one would suspect at first glance might possess the required property, but one easily verifies on closer inspection that it does.

Thirdly, in our investigation of series of numbers which may or may not go on forever, we shall probe the phenomenon of the *persistence* of numbers. The concept of persistence has been defined as the number of steps necessary to reduce a number to a single digit by multiplying all its digits together to obtain a second number, and then multiplying all the digits of that number to obtain a third number, and so on, until a single digit number is obtained. As an example, the integer 679 has a persistence of 5 since

$$
\begin{array}{llll}
1) & 6\times7\times9 & = & 378 \\
2) & 3\times7\times8 & = & 168 \\
3) & 1\times6\times8 & = & 48 \\
4) & 4\times8 & = & 32 \\
5) & 3\times2 & = & 6
\end{array}
$$

Not only does 679 have a persistence of 5, it is in fact the very smallest number which does have a persistence as large as 5. That is to say, all numbers less than 679 can be reduced to a single digit by the method set out above in 4 steps or less. It is not true, of course, that all numbers larger than 679 have a persistence of 5 or greater. Quite obviously arbitrarily large numbers can be written which have a persistence of only 1. Any number, for example, which contains a zero as one of its digits, produces the single digit resultant 0 after only one step. Nevertheless the sequence of integers formed by the smallest numbers of persistence 1, 2, 3, 4, 5, ... does appear to be a uniformly increasing series. Let us investigate how it begins.

With a little thought it is clear that the smallest number of persistence 1 is 10, while the smallest number of persistence 2 is 25 (which generates the sequence $25\rightarrow10\rightarrow0$). A simple check of integers less than 100 quickly produces the smallest numbers of persistence 3 and 4. They are 39 and 77 respectively, and generate the persistence sequences $39\rightarrow27\rightarrow18\rightarrow9$ and $77\rightarrow49\rightarrow36\rightarrow18\rightarrow8$. Our series of interest therefore begins

$$10, 25, 39, 77, 679, ...$$

Once again we ask 'does it go on forever?' The answer is not at all obvious and, in fact, at the present time is unknown. Nevertheless, with the aid of a computer more terms in the series have been discovered. At the time of writing some 11 terms in the series have been generated; they are the above 5 followed by

6,788,   68,889,   2,677,889,   26,888,999,
3,778,888,999   and   277,777,788,888,899.

We note that not only are the terms getting bigger, but the ratio between successive terms is also growing. The last term, in particular, is more than seventy thousand times its predecessor, while the tenth term is only some one hundred and forty times its predecessor. The earlier terms are more typically a factor of ten or less larger than the terms which they immediately follow.

There is therefore a suggestion that the series may be about to 'blow up'. By this we mean that one of the next few terms (and possibly the very next term itself) could be infinite, thus terminating the sequence. This is a rather new way of ending a series and would imply that there exists a largest persistence of all. Such an idea comes as a surprise to most, since there is a tendency to feel that if one can only make a number sufficiently long and avoid zeroes then persistencies of any magnitude should be obtainable. If true, however, it means that all numbers, *no matter how large,* have a persistence less than or equal to some finite value like 25, or 15, or just possibly even 11. A computer search for a number with persistence greater than 11 has been carried out all the way up to 10^{50} without success. This means that the next term in the persistence series has at least 50 digits, and is consequently at least a factor of 10^{35} bigger than its predecessor. This is without doubt an enormous gap but, as we have so often stressed before, infinity is very far off and a computer is never going to reach it. I find this a satisfying state of affairs. It means that if the persistence series does end then it will require a mathematical proof and not just computer money to establish it as fact. For the moment the sentiments of number theorists favor a termination but, until the proof appears, doubts must always remain.

I shall finish this chapter with brief reference to a slightly different, but related, class of unsolved sequence problems called 'looping problems'. They also involve series of integers generated according to a rule, but this time one asks whether the series always enter one or more cycles (or 'loops') in which a finite set of integers keeps cycling forever. Perhaps the simplest example of this is a sequence which starts with any integer and halves it if it is even, or triples it and adds one if it is odd. Repeating this recipe generates a never ending sequence of integers. Let us investigate by looking at the simplest starting numbers 1,2,3, etc. We find the respective sequences:

$$1, 4, 2, 1, 4, 2, 1, 4, 2, \ldots$$
$$2, 1, 4, 2, 1, 4, 2, 1, 4, 2, \ldots$$
$$3, 10, 5, 16, 8, 4, 2, 1, 4, 2, \ldots$$

which all quickly enter the same 142142142 loop. Let us try one more, starting (say) with 7:

$$7, 22, 11, 34, 17, 52, 26, 13, 40, 20,$$
$$10, 5, 16, 8, 4, 2, 1, 4, 2, 1, 4, 2, \ldots$$

This time it takes a little longer, but again the 142-loop is reached in the end. The question is 'must this 142-loop always eventually appear?' Well, computers have checked out all starting integers up to 60 million without finding a single exception. This is very suggestive, but no proof has yet been given that the 142-loop must necessarily be the final demise of all such series. Could there possibly be a starting integer which generates a non-looping sequence which gradually increases forever? At present no-one knows.

By changing the recipe for adding terms (for example by replacing the 'triple and add one' rule to 'triple and subtract one') other sequences with other looping endpoints can be formed with ease. You might like to try out a few for yourselves. I am sure that very little is known for sure about any of them, so that your own research may easily lead to new conjectures and yet more headaches for the number theorist.

11. FERMAT'S LEGENDARY LAST THEOREM

Among the vast population of non-mathematicians and non-scientists of this world perhaps the two best known examples of algebraic equations are the $E=mc^2$ of Einstein and the $a^2+b^2=c^2$ of Pythagoras. The Pythagorean theorem, as it is more usually known, is a statement about right-angled triangles which says that the square of the longest side c of such a triangle is equal to the sum of the squares of the two shorter sides a and b. Although the original statement was basically geometric it did pose an interesting associated mathematical question, namely 'what are the whole number solutions of the Pythagorean equation?' Of course, there is no particular reason why one should restrict solutions of

$$a^2 + b^2 = c^2$$

to whole numbers, and in general one can always find a real number c to satisfy the equation when a and b have been given. But mathematicians have always found something fascinating about whole numbers and in the present context they provide added interest to the equation, since only rather special sets of three integers will satisfy it. One solution has been known since early Egyptian times, namely

$$a = 3, \quad b = 4, \quad c = 5.$$

Indeed the Egyptians built their pyramids by marking a rope into three, four, and five units so that it fell automatically into the form of a right angled triangle. Whole numbers, you see, are so convenient to measure with. Other small integer solutions are not too difficult to find, such as

$$a = 5, \quad b = 12, \quad c = 13,$$

but the more general problem concerning all the whole number solutions is obviously not a trivial one. In a wider context it is an example of a larger class of problems involving the solution of algebraic equations in terms of integers. The general field is termed Diophantine analysis after a prominent Greek mathematician Diophantus of Alexandria who probably lived in the third century A.D.

Diophantus left behind him a little book which contained many problems asking for solutions to simple algebraic equations in terms of whole numbers. Today Diophantine analysis is a rather well-developed but extraordinarily difficult branch of number theory. The difficulties arise because there are no known general methods for solving problems of this kind if the relevant equations contain powers larger than or equal to 2. Although the general solution to the particular Pythagorean problem cited above is known, and has in fact been known for a long time, only slightly more complex variations still defy solution. One well-known problem in the latter category concerns the question of finding the size of a building-brick whose three edges, three face diagonals and single body diagonal (from one corner to the farthest opposite corner) are all whole numbers. In this problem there are 7 unknowns connected by four equations (all of the simple Pythagorean form) with each equation connecting three of the unknowns. Like so many other Diophantine problems it has never been shown to be impossible, nor has any solution ever been found, even with the aid of the largest computers. It follows that there may be no such brick. On the other hand there may be many, or even an infinite number all different. No-one knows. But let us continue our major story.

Pierre de Fermat was a seventeenth century amateur number theorist and was about thirty years old when a copy of Diophantus' work fell into his hands. Working through the many problems contained in the book, Fermat often went beyond the requested solutions if he saw an extension or a generalization which interested him. He would often scribble notes in the margin of his copy of Diophantus, and it is in this connection that we first meet the legendary 'last theorem'. Fermat, in the margin next to Problem 8 in Book 2 which concerned the whole number solutions to the Pythagorean equation, added a note in Latin. He wrote that an equation like $a^2+b^2=c^2$, although it has an infinite number of whole number solutions when the powers are equal to 2 as shown, has *no solutions whatsoever* if the powers are cubes, or equal to 4, 5, 6, or *any* integer greater than 2. He continued "I have discovered a truly marvellous demonstration which this margin is too narrow to contain". It has since been said that the width of that margin has affected the entire subsequent development of number theory.

In any discussion of this marginal note, let it first be made clear that the absence of a proof in this context was not at all unusual for Fermat. He left behind many other examples and for a lesser man not too much attention would be paid to them, because a theorem without a proof is of limited value. But Fermat was a man of unimpeachable integrity and was perhaps the greatest number theorist who has ever lived. In every other case when he said he had proven a theorem, even though the proof was never found in the writings he left behind, the correctness of the theorem was eventually established by other mathematicians at a later (and often very much later) date. On the other hand, when he said he merely had a hunch about something but could not prove it, then sometimes these hunches turned out to be correct and sometimes they didn't. Among the latter is a very famous example which we have already met; namely Fermat's suspicion that numbers of the form

$$2^{2^n} + 1$$

are always prime. They are not; but Fermat never claimed that he could *prove* that they were. What he did say in the Diophantine context was that he could prove that any equation of the form

$$a^n + b^n = c^n,$$

where n is an integer greater than 2, has no solution at all in which a, b, c are whole numbers. Only this theorem, among all those which Fermat claimed he could prove, still remains unproven by all the generations of mathematicians who have followed. It has therefore come to be called Fermat's 'last theorem' and remains one of the greatest unsolved problems of modern mathematics.

To prove the statement is evidently very difficult; so difficult that many mathematicians felt that Fermat must have been mistaken when he thought that he had found a proof. They are not saying that the statement is necessarily incorrect; just that the mathematical tools and techniques necessary for attacking the problem just could not have been available to Fermat. But did Fermat really have a 'marvellous demonstration'? It is just possible that he did, but in references made elsewhere Fermat only ever refers to restricted forms of the statement. For example he does say in other places that he could show that

$$a^3 + b^3 = c^3$$

and

$$a^4 + b^4 = c^4$$

have no solutions in integers. These more restricted forms of the last theorem have been proven by others. If Fermat really could prove that the statement was true for *all* powers greater than 2, would he not have mentioned that fact here too? Maybe he thought that he had a completely general proof when he made his marginal note, but later found an error in the argument. Since his marginal notes were never intended for publication, he would have no particular reason to go back to his book and locate and amend the note. We shall never know. In any case the general theorem has yet to be either proved or disproved and this is certainly not through want of effort. In fact, the study of this theorem has prompted some of the greatest of all creations of mathematical thought.

How does one go about proving anything concerning numbers which can be arbitrarily large? Although one could conceivably *dis*prove Fermat's last theorem with a single counter-example (that is finding an actual set of integers a, b, and c for which the equation $a^n + b^n = c^n$ is satisfied for some value of n greater than 2), and this might be achieved with a computer by simply testing all the numbers up to a certain limit, one can never prove that the theorem is true by this method. No computer, even a futuristic one with undreamed of efficiency, can ever test the entire infinity of counting numbers. Given this fact it is easy for non-mathematicians to quickly become discouraged and feel that it is impossible to do. However, Fermat himself left behind a proof for $n = 4$, albeit in a somewhat disguised form concerned directly with another problem. The method consists of assuming at the outset that the equation $a^4 + b^4 = c^4$ *does* have a solution in whole numbers, and then finding a construction to obtain a smaller set of numbers which also satisfies the same equation. If this can be done, and it can for $n = 4$, then we can go on repeating this procedure to smaller and smaller numbers forever. Since the solution involves only whole numbers, this necessarily leads to a contradiction and hence shows that the original assumption (namely that $a^4 + b^4 = c^4$ had a whole number solution) must be incorrect. The technique, for obvious reasons, is called the method of infinite descent.

We can now make a further observation. It is that once the theorem is proved for $n = 4$ then it is trivial to establish it for any multiple of four (that is for $n = 8$, 12, 16, 20, 24, ... and so on). Even we amateurs can see this because, for example, $a^8 + b^8 = c^8$ can be rewritten as

$$(a^2)^4 + (b^2)^4 = (c^2)^4,$$

which is just a particular example of the fourth power result over again. This same trick can be used in a wider context to establish that if Fermat's last theorem is valid for any particular value $n = m$ of the exponents, then it is also true for

$$n = 2m, 3m, 4m, 5m, \cdots$$

all the way to infinity. If you think a bit about this, it means that one
has only to worry about the theorem when the exponent n is equal to a
prime number greater than or equal to 3.

Among his other papers, Fermat also made a specific separate
statement that he could prove the theorem for $n=3$. Once again no
actual example of the proof was ever found in the papers he left
behind. Nevertheless, in this particular case, the proof was rederived
about one hundred years after Fermat's death (which occurred in 1665)
by Euler, again using the method of infinite descent. It is perhaps
comforting to those of us who have no delusions of grandeur when it
comes to mathematics, to learn that even the great Euler could make
an error. His original proof for $n=3$ did in fact contain a serious flaw,
but one which could be and was eventually overcome. The proof for
$n=5$ followed in the year 1820 and that for $n=7$ some 15 years later.
Unfortunately these proofs were rapidly becoming longer and more
difficult, and future progress seemed unlikely unless a major
breakthrough could be accomplished in the basic method used. With
this object in mind many efforts were made and some capable
mathematicians announced such breakthroughs from time to time, only
to be embarrassed later by having their errors publicly demonstrated.

It was left to a German number theorist named Ernst Kummer in
1847 to announce what amounts to the greatest advance ever made on
Fermat's last theorem. The proofs for $n=5$ and $n=7$, which had been
announced only a decade or two earlier, were rightly considered to be
great accomplishments, but Kummer was able to prove the theorem
true for all prime numbers up to and including the power $n=31$. In
fact he came very close to proving the theorem true for all prime
number exponents with n less than 100, since only the values $n=37$,
59 and 67 eluded his method among the two-digit primes. The great
power of Kummer's method was that it could establish a certain
condition which, if obeyed by a particular prime number p, made the
solution of $a^p + b^p = c^p$ in whole numbers an impossibility. Today the
prime numbers satisfying this condition, for which Fermat's last
theorem is thereby established by this method, are said to be 'regular
primes'.

Kummer thought initially that the number of regular prime numbers
in the above sense was infinite, but he soon realized that he could not
prove it. Such a proof is still missing although mathematicians feel

sure (or at least as sure as they can about anything which still lacks a proof) that they are. About 60% of all the prime numbers within the range of the largest computers today are regular, and there are good reasons for supposing that the majority of all primes (right up to infinity) are. Thus, in a single major work, Kummer extended the proof of Fermat's last theorem from the prime exponents 3, 5, and 7 to a probably infinite set of prime numbers perhaps containing more than half of all the primes which exist.

In his later years Kummer further liberalized the restrictions on the primes which satisfied the last theorem and managed to include the irregular prime numbers 37, 59, and 67. Subsequently Kummer's techniques were extended and improved still further until today all primes within the range of even the largest computers are exponents for which the last theorem is known to hold (at this writing the limit is now well beyond $n = 100,000$). Nevertheless it is still possible, no matter how unlikely, that Fermat's last theorem is true for only a finite number of prime exponents, and that there exists some extremely large prime number p above which the theorem breaks down, perhaps even for *all* larger prime numbers. This is certainly extremely unlikely, but not impossible. What these computations do show is that a counter-example to Fermat's last theorem would necessarily consist of numbers so large that testing them would be far beyond the capabilities of even today's fastest computers. Specifically it has been shown that a counter-example would now have to involve numbers of more than one million digits.

In a practical sense, therefore, the infamous last theorem is valid. On the other hand, in the true mathematical sense it remains very much unproven. It still appears that another fundamental breakthrough will be necessary before significant further progress can be made. Since the existing partial proofs are already so complicated that few non-specialists can hope to understand them, the future does not look very bright and little work by professional number theorists now seems to be carried out on the subject. But what if Fermat really did have that 'marvellous demonstration' which his marginal note claimed. If he did, it must necessarily have been a reasonably elementary one, and it may just possibly still be out there waiting to be rediscovered. It is just such a belief which has sparked the efforts of amateurs for more than 300 years since Fermat pondered the problem. Surely the story should have a happy ending. But is it possible that a problem which has beaten the best professional mathematicians for so long could actually have an elementary proof? We can always hope so, and in Appendix 3 (in order to bolster the faith) I give an example of a problem from a somewhat different (but not altogether unrelated) context which also defeated the best minds for over 100 years but whose proof, when finally found, was so simple that even children could understand it. Turn to Appendix 3 if you are interested.

We have referred to the fact that it is possible to *dis*prove any theorem of the Fermat type by finding a single counter-example. Such a demonstration would, of course, be extremely convincing once it had been discovered. A nice example of this kind does exist in number theory, and concerns an extension of Fermat's last theorem which was suggested initially, I believe, by Euler. This extension proposed that equations of the kind

$$a^n + b^n + c^n + d^n + ... = z^n,$$

in which all of a, b, c, d, ..., z and n are whole numbers, have no solutions at all with less than n (but of course more than 1) terms on the left-hand side. For $n=3$ this would mean that there are no whole number solutions to

$$a^3 + b^3 = c^3$$

and, as we have seen, this is a particular case of Fermat's last theorem which was proved by Euler himself. On the other hand there are many known solutions in integers to

$$a^3 + b^3 + c^3 = d^3,$$

the one involving the smallest numbers being

$$3^3 + 4^3 + 5^3 = 6^3.$$

This is a beautiful kind of extension of the well known 3, 4, 5, Pythagorean triangle result. Unfortunately obvious further extensions do not persist to higher powers, and the sum of the fourth powers of 3, 4, 5, and 6, for example, does not equal the fourth power of 7. For $n=4$ the Euler extension of Fermat's last theorem implies that there are no whole number solutions to

$$a^4 + b^4 + c^4 = d^4$$

and indeed none is yet known. Once again, however, solutions are known for the slightly expanded form

$$a^4 + b^4 + c^4 + d^4 = e^4,$$

the simplest being

$$30^4 + 120^4 + 315^4 + 272^4 = 353^4.$$

Proceeding to $n = 5$ the Euler assertion states that there are no whole number solutions to

$$a^5 + b^5 + c^5 + d^5 = e^5$$

and it is with regard to this last equation that a computer search in 1966 found the single exception needed to disprove the whole speculation. This rather noteworthy exception is, for the record

$$27^5 + 84^5 + 110^5 + 133^5 = 144^5.$$

Finally, before leaving relations of this type altogether, let us return one more time to the starting point of the whole venture; namely the Pythagorean relationship $3^2 + 4^2 = 5^2$. This equation is unique among the $a^2 + b^2 = c^2$ group because of the pleasing consecutive integer sequence 3, 4, 5. It is not difficult to show that no other set of consecutive integers can satisfy this equation, but one can perhaps ask whether it is possible to retain a consecutive pair of integers just on the left hand side and find additional solutions. It happens that there are other solutions of this slightly restricted kind, two examples being

$$20^2 + 21^2 = 29^2$$
$$119^2 + 120^2 = 169^2.$$

Moreover there are an infinite number of solutions like this and it is tempting to ask whether one can go one step further to obtain forms like $a^2 + b^2 + c^2 = d^2$ in which the integers a, b, and c are consecutive whole numbers. Surprisingly, perhaps, no equations of this kind (with three consecutive integers on the left hand side) exist. Similarly it is known that none exists with 4, or 5, or 6 consecutive integers on the left hand side either. But for the persistent investigator there is a reward, for there are other solutions out there somewhere. The next simplest one is found when there are no less than 11 consecutive integers on the left hand side. It is

$$18^2 + 19^2 + \cdots + 27^2 + 28^2 = 77^2.$$

What is more, many other solutions are known with 11 terms on the left hand side, one being

$$38^2 + 39^2 + \cdots + 47^2 + 48^2 = 143^2.$$

It is now interesting to ask whether there might even be an infinite number of solutions with 11 consecutive terms on the left hand side of the equation, and it turns out that there are. But the story does not end there. It has been established that equations of this type, where n consecutive squares added together on the left hand side equal a single perfect square on the right hand side, can have solutions only for rather special values of n of which the first two are 2 and 11. The complete set of allowed values of n less than 100 is

$$2, 11, 23, 24, 26, 33, 47, 49$$

$$50, 59, 73, 74, 88, 96, 97.$$

I think you will agree that this is a most unlikely looking set of numbers. Can you imagine an I.Q. test which included the question 'continue the series 2, 11, 23, 24, 26, ...'. It would be more than a little unfair, but at least *we* now know the answer. Beyond 100 about one tenth of the possible n-values allow solutions. A few examples of the more impressive looking relationships with larger n are

$n = 184$: $7^2 + 8^2 + 9^2 + \cdots + 188^2 + 189^2 + 190^2 \quad = \quad 1518^2$

$n = 289$: $20^2 + 21^2 + 22^2 + \cdots + 306^2 + 307^2 + 308^2 \quad = \quad 3128^2$

$n = 458$: $1081^2 + 1082^2 + \cdots + 1537^2 + 1538^2 \quad = \quad 28167^2.$

It seems that there are an infinite number of solutions for each allowed value of n except when n itself is a perfect square (for example the cases $n = 49$ and $n = 289$ above) in which case only a finite number of solutions exists.

12. SHAPELY NUMBERS AND MR. WARING

Ancient Greek mathematicians, and especially the followers of Pythagoras and his school, were entranced by numbers which could be made up by arranging points in regular patterns on a plane or in space. We are already familiar with the 'square numbers' (i.e. the squares of the integers) beginning 1, 4, 9, 16, 25, ... and these particular numbers can be given a picturesque representation by building up bigger and bigger squares on a plane in the following manner;

```
*   * *   * * *   * * * *   * * * * *
    * *   * * *   * * * *   * * * * *
          * * *   * * * *   * * * * *
                  * * * *   * * * * *
                            * * * * *
```

Since a square is only one of an infinite number of regular planar figures (and by regular we mean figures with all angles equal and all side-lengths equal), it is clear that other groups of numbers can be made up by using these other figures as foundations.

The simplest such figure, with only three equal angles and three equal sides, is the equilateral triangle. Using it as a basic building block in the following manner;

```
*     *     *       *         *
    * *   * *     * *       * *
        * * *   * * *     * * *
              * * * *   * * * *
                      * * * * *
```

we find that it generates the numbers

1, 3, 6, 10, 15, 21, 28, 36, 45, 55, 66, ...

which are consequently referred to as the *triangular* numbers. After the triangular numbers and the squares it is now straightforward to extend the idea to those sets of integers which can be generated by regular 5-sided figures (called pentagons), 6-sided figures (hexagons), and so on. The general building sequences are shown in Figure 5 and, for the *pentagonal* numbers, generate a series beginning

1, 5, 12, 22, 35, 51, 70, 92, ...

and for the *hexagonal* numbers

$$1, 6, 15, 28, 45, 66, 91, \ldots$$

We see that the triangular numbers are formed by the successive addition of every integer starting with 1. By this we mean that they are generated by the summation sequence

$$
\begin{aligned}
1 &= 1 \\
1+2 &= 3 \\
1+2+3 &= 6 \\
1+2+3+4 &= 10 \\
1+2+3+4+5 &= 15 \\
1+2+3+4+5+6 &= 21 \\
1+2+3+4+5+6+7 &= 28
\end{aligned}
$$

and so on. The squares, in turn, can be formed in a similar way by adding in sequence all the *second* integers following 1 (or equivalently all the odd numbers) as follows:

$$
\begin{aligned}
1 &= 1 \\
1+3 &= 4 \\
1+3+5 &= 9 \\
1+3+5+7 &= 16 \\
1+3+5+7+9 &= 25 \\
1+3+5+7+9+11 &= 36 \\
1+3+5+7+9+11+13 &= 49
\end{aligned}
$$

The pentagonal and hexagonal numbers follow the same sort of pattern, which is made clear in Figure 5, and are formed by successive summation of all third and all fourth numbers after 1 respectively. This gives for pentagonal numbers the sequence

$$
\begin{aligned}
1 &= 1 \\
1+4 &= 5 \\
1+4+7 &= 12 \\
1+4+7+10 &= 22 \\
1+4+7+10+13 &= 35 \\
1+4+7+10+13+16 &= 51 \\
1+4+7+10+13+16+19 &= 70
\end{aligned}
$$

and for hexagonals the sequence

$$
\begin{aligned}
1 &= 1 \\
1+5 &= 6 \\
1+5+9 &= 15 \\
1+5+9+13 &= 28 \\
1+5+9+13+17 &= 45 \\
1+5+9+13+17+21 &= 66
\end{aligned}
$$

All these sequences can naturally be continued indefinitely to arbitrarily large numbers and, like many after them who looked at these relationships, the early Greeks found the shapes and inter-relationships of these 'polygonal' numbers extremely interesting.

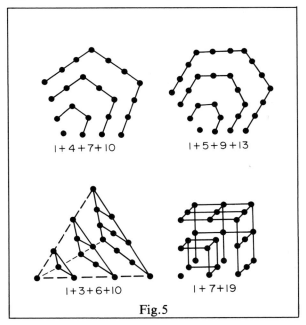

Fig.5

Firstly you notice that the same number can arise in more than one sequence. Just among the very small numbers we see that 66 is both a triangular number and a hexagonal number, while 36 is both a triangular number and a square. All sorts of questions can be asked concerning polygonal numbers, some of fundamental significance and others perhaps less so. One immediate observation is that two repeating digit numbers 55 and 66 appear among the early terms in the triangular number sequence. Does this perhaps suggest that repeating digit integers are relatively common among the triangular numbers? It most certainly does not. Except for 666, no other repeating digit number of any length is a triangular number. We are already forewarned that appearances can be deceptive. What about those numbers which are both triangular and square? Are there many of these? This question seems first to have been asked by Euler who proceeded to find a general formula for such numbers. From the summation sequence for triangular numbers we note that the nth triangular number is the sum of the series

$$1 + 2 + 3 + \cdots + (n-2) + (n-1) + n.$$

By adding the terms in pairs from each end (that is the first with the last, the second with the one from last, and so on) this sum can be thought of as $\frac{1}{2}n$ lots of $(n+1)$. This means $\frac{1}{2}n$ times $(n+1)$ and in algebraic notation is written as $\frac{1}{2}n(n+1)$. If this general triangular number is also equal to a perfect square, say m^2, then

$$\frac{1}{2}n(n+1) = m^2$$

must be the defining equation for 'triangular-square' numbers. It was therefore Euler's task to solve this equation in whole numbers. In the language of the last chapter it is a typical Diophantine problem and hence requires techniques which are surprisingly difficult. Nevertheless Euler was a match for it and he established that there are an infinite number of solutions which begin

$$1, \quad 36, \quad 1{,}225, \quad 41{,}616, \quad 1{,}413{,}721,$$
$$48{,}024{,}900, \quad 1{,}631{,}432{,}881, \quad \ldots$$

but evidently become large very quickly. As a result not many of them are likely to be found by the simple trial technique of writing down all the triangular numbers and squares up to a manageable limit and looking for coincidences. A glance at the striking simplicity of the defining Diophantine equation $\frac{1}{2}n(n+1) = m^2$, and at the sheer size of even some of the smaller solutions, gives one an excellent appreciation of the difficulties which these kinds of problems possess.

On a more general level, it is easy to show that the triangular numbers and the square numbers are related in a strikingly simple manner. It is that the sum of any two consecutive triangular numbers is always a square number and, moreover, *all* square numbers can be formed in this way. For example

$$1 + 3 = 2^2$$
$$3 + 6 = 3^2$$
$$6 + 10 = 4^2$$
$$10 + 15 = 5^2$$

and so on. These relationships were known to the early Greeks and later-on Fermat himself expressed an interest in general relationships between the shapely numbers. He wrote (yes, it was once more in the margin of his copy of Diophantus) that *every* integer is either triangular or is the sum of 2 or 3 triangular numbers. Similarly, every integer is either a square or is the sum of 2, 3, or 4 squares; is either pentagonal or is the sum of 2, 3, 4, or 5 pentagonal numbers, and so forth. Once again he did not leave for posterity any proof of these statements, and it remained for other researchers over the following centuries to provide the details. The proof for squares was given by the French mathematician and astronomer Joseph-Louis Lagrange, that for triangular numbers by Gauss, and finally another French mathematician Augustin Cauchy proved the general case. It is in connection with deceptively simple statements of this kind that we shall shortly meet Mr. Waring, who will lead us into associated speculations which have puzzled the best mathematical minds for two centuries.

Triangular numbers often appear in the most unexpected places. Firstly they give directly the number of ways in which two objects can be chosen from a larger collection. For example, there is 1 way of choosing two objects from a set of 2 (if we ignore the order in which they are picked), 3 ways of choosing them from a set of 3, 6 ways from a set of 4, 10 ways from a set of 5, and so on. More surprisingly, all perfect numbers are known to be triangular. This means, among other things, that a perfect number of pool or billiard balls can always be 'racked up' in a triangle of a suitable size; even if we have $2^{44496} \times (2^{44497} - 1)$ of them. Even those triangular-square numbers which we wrote out above turn out to be useful. They can be used to generate right-angled (or Pythagorean) triangles of a particular kind; namely those in which the two shorter sides differ in length from each other by a single unit. The smallest such triangle is the well-known 3, 4, 5 one. The next smallest are those with side lengths 20, 21, 29 and 119, 120, 169. Using the triangular-square numbers (in a manner too complex to be detailed here) triangles of this kind can be generated with side lengths up to astronomical dimensions. For example, the one hundredth such triangle has two shorter sides which are each expressed by 77-digit numbers (and which, of course, differ by 1). Although no right-angled triangle can ever be isosceles (that is have two sides of *exactly* equal length) this giant, it has been pointed out, would need some very close scrutiny to establish for certain that it was not equal sided. As a practical matter, if each of its smaller sides were a million light years long (where a light year is the distance travelled by light in one year) they would differ in length by an amount smaller than the diameter of an atomic nucleus.

Shapely numbers, of course, are not necessarily restricted to planar (that is two-dimensional) figures. One can just as easily build up number sequences from regular solid figures in three dimensions. The regular solid with the smallest number of faces is the tetrahedron, which is a pyramid with a three-sided base (having all its faces equilateral triangles). This solid enables us to build up the so-called *tetrahedral* numbers (see Figure 5) as follows:

$$1, 4, 10, 20, 35, 56, ...$$

from which we see that they are just the sum of triangular numbers starting from 1 in the form

$$1, \ 1+3, \ 1+3+6, \ 1+3+6+10, \ 1+3+6+10+15, \ ....$$

These tetrahedral numbers measure the different ways of choosing three objects from a larger group. Thus there is 1 way of choosing three objects from 3, 4 ways of choosing them from 4, 10 ways of choosing them from 5, and so on. Going beyond the tetrahedron the next simplest regular solid is the cube, which is perhaps the most familiar solid of any kind. It gives rise to the cubic numbers (or cubes)

$$1, 8, 27, 64, 125, 216, ...$$

and they can be related to the planar numbers discussed above in several interesting ways. They can, for example, always be formed from the odd integers by the following summation procedure

$$
\begin{aligned}
1^3 &= 1 \\
2^3 &= 3 + 5 \\
3^3 &= 7 + 9 + 11 \\
4^3 &= 13 + 15 + 17 + 19 \\
5^3 &= 21 + 23 + 25 + 27 + 29
\end{aligned}
$$

a series which compares interestingly with the way in which square numbers are also made up from summations involving the same odd numbers, namely

$$1^2 = 1$$
$$2^2 = 1 + 3$$
$$3^2 = 1 + 3 + 5$$
$$4^2 = 1 + 3 + 5 + 7$$
$$5^2 = 1 + 3 + 5 + 7 + 9.$$

These two simple patterns now enable us to see something quite profound in the relationship between squares and cubes - a result which often astounds students when they first meet it. The sum of all the cubes

$$1^3 + 2^3 + 3^3 + 4^3 + \ \ldots\ldots$$

up to any arbitrary endpoint must, from the first pattern, just be the sum of all the odd numbers up to some value. But, from the second pattern (for squares), the sum of all the odd numbers up to any point must always be an exact square. It follows that the sum of all the cubes up to any limit must always be an exact square. But we can go further. From the first (cubes) pattern, the sum of all the cubes up to (say) 4^3 is

$$1 + (3 + 5) + (7 + 9 + 11) + (13 + 15 + 17 + 19)$$

and is the sum of $1+2+3+4 = 10$ consecutive odd numbers starting from 1. From the second (squares) pattern we see that any square n^2 is the sum of n consecutive odd numbers starting from 1. It follows that we must have

$$1^3 + 2^3 + 3^3 + 4^3 = (1+2+3+4)^2 = 10^2.$$

We can generalize this in an obvious way to conclude that the sum of all the cubes from 1^3 to n^3 is just the square of the sum of the numbers from 1 to n. In symbols this is

$$1^3 + 2^3 + 3^3 + \cdots + n^3 = (1+2+3+\ldots+n)^2 .$$

Since we have already found an expression for the sum of the numbers from 1 to n in connection with our discussion of triangular numbers (namely $\frac{1}{2}n(n+1)$) we conclude finally that

$$1^3 + 2^3 + 3^3 + \cdots + n^3 = [\tfrac{1}{2}n(n+1)]^2 .$$

For example, the sum of all the cubes from 1^3 to 100^3 is equal to

$$[\tfrac{1}{2} \times 100 \times 101]^2 = 5050^2 = 25{,}502{,}500$$

which is certainly quicker than doing it the hard way! We learn from this that a little attention to number patterns and shapes will often enable us to deduce number relationships which are not at all obvious at first glance.

Apart from the cube, only three more fully regular solids exist; the octahedron (with 8 identical faces), the dodecahedron (with 12) and the icosahedron (with 20). Each may be used to generate shapely numbers with their own character. More generally still, in both the planar and the solid contexts, it is not necessary to restrict attention to fully regular figures in order to produce shapely number sequences. Other shapes such as stars and square pyramids will do just as well, and the possibilities for research are virtually boundless.

Now what about Mr. Waring? Edward Waring was a rather modest English mathematician of the eighteenth century whose interest in the connection between shapely numbers (particularly squares and cubes) and integers led him in 1770 to make the statement that "every positive integer can be made up as the sum of no more than 4 squares, no more than 9 cubes, no more than 19 fourth powers, and so on". The assertion, now named after Waring, was not presented with any accompanying grand theory, but was probably just a plausible surmise on Waring's part arrived at by examining the smaller integers. We have already referred to the conjecture for squares, which had been made much earlier by Fermat, in connection with a more general statement concerning all the planar shapely numbers. Waring's contribution was therefore to extend this idea to solid shapely numbers and in particular to cubes. But first let us see specifically how it works for squares. In Table 8 we show how the integers from 1 to 20 can be made up from four or less squares. From this short list it appears that 4 squares will suffice and that all four are not needed very frequently. Proceeding in an similar fashion with cubes one quickly finds that 9 cubes are required for the number 23 in the form

$$23 = 2^3 + 2^3 + 1^3 + 1^3 + 1^3 + 1^3 + 1^3 + 1^3 + 1^3$$

but that all the rest, at least up to 100 (which are easily tested), can be done with less.

TABLE 8		
The first 20 integers expressed as the sum of the smallest possible number of exact squares		
1	=	1^2
2	=	$1^2 + 1^2$
3	=	$1^2 + 1^2 + 1^2$
4	=	2^2
5	=	$2^2 + 1^2$
6	=	$2^2 + 1^2 + 1^2$
7	=	$2^2 + 1^2 + 1^2 + 1^2$
8	=	$2^2 + 2^2$
9	=	3^2
10	=	$3^2 + 1^2$
11	=	$3^2 + 1^2 + 1^2$
12	=	$3^2 + 1^2 + 1^2 + 1^2$
13	=	$3^2 + 2^2$
14	=	$3^2 + 2^2 + 1^2$
15	=	$3^2 + 2^2 + 1^2 + 1^2$
16	=	4^2
17	=	$4^2 + 1^2$
18	=	$3^2 + 3^2$
19	=	$3^2 + 3^2 + 1^2$
20	=	$4^2 + 2^2$

As previously mentioned the fact that 4 squares would suffice for *any* finite integer was first proved by Joseph-Louis Lagrange, and followed soon after Waring's conjectures. The proof of the sufficiency of 9 cubes, however, turned out to be a much more difficult problem and was not solved until the early years of the 20th century. Nevertheless, a proof was eventually given, and that might have been the end of the story for cubes except for one puzzling fact; it was that with the exception of the numbers 23 and 239, where

$$239 = 5^3 + 3^3 + 3^3 + 3^3 + 2^3 + 2^3 + 2^3 + 2^3 + 1^3$$

no other integers could be found which required the full 9 terms.

In 1939 the astonishing proof was presented that 23 and 239 are, in fact, the only two numbers among the complete infinity of integers which require the sum of 9 cubes to produce them. All the rest can be obtained using 8 cubes or less. Put in another way, all the integers greater than 239 can be expressed as the sum of 8 or fewer cubes. But now comes the interesting challenge. How are we to know that there isn't some higher number beyond which only 7 are needed, or even 6, or 5? This new slant on the old Waring problem set into motion further number research and eventually it was shown that only 15 integers require the sum of 8 cubes; they are

$$15, 22, 50, 114, 167, 175, 186, 212,$$
$$213, 238, 303, 364, 420, 428, 454.$$

It follows that all integers larger than 454 can be formed by summing *seven* or fewer cubes. But how far can this sort of thing go on? There must certainly be some number, 7 or less, which is required even at the highest bounds.

The complete answer to the remaining riddle is still not known, although it is now quite widely believed that this limiting number, 'big gee' as it is sometimes called, is actually less than 7. It *is* known to be at least 4, so that the remaining candidates are

$$\text{big gee} = 4, 5, 6, \text{ or } 7.$$

Back in 1851 the German mathematician Karl Jacobi had tabulated all the numbers to 12,000, writing each as the sum of as few cubes as possible. He found that 121 numbers up to 12,000 required 7 cubes but that only 3 of them, namely 5,306, 5,818, and 8,042, were larger than five thousand. On the basis of the increasingly large gaps between these numbers Jacobi concluded that the largest number requiring as many as 7 cubes might well be the largest which he had found, namely

$$8,042 = 19^3+10^3+4^3+4^3+3^3+3^3+1^3.$$

Even to the present day, using the fastest available computers, no larger number requiring as many as 7 cubes has been found, but whether 'big gee' is truly 4, 5, 6, or just possibly even 7, is still not known. It *is* known that every integer greater than 12,758 can be represented as the sum of *distinct* cubes (that is without any particular cube being repeated). It is also known that all integers greater than 128 can be represented as the sum of distinct squares.

The general situation for squares is not so exciting as that for cubes. As mentioned before it has been proven that 4 squares are needed to express all the integers as the sum of squares. In addition it is also known that there is no large finite integer above which less than 4 will do. Nevertheless it is possible to extend Waring's problem for cubes in a number of directions involving other shapely numbers, and even to spaces with more than three dimensions. In this respect most recent research has followed the extension to powers higher than cubes, that is to sums involving terms n^k in which the exponent $k = 4$, 5, 6, ... and so on. In shapely language, numbers like n^k with k greater than 3 refer to 'hypercubes' in the higher dimensions, although this more difficult picture-concept need be of no concern to us here. Waring himself considered this extension and implied that the number of terms required for $k = 4$ was 19, and that for any value of k would always be finite. This was eventually proved in the year 1909 by another prominent German mathematician David Hilbert. For these higher powers, at least up to powers $k = 20,000$ and quite possibly for all k, the number of terms required is given by a formula first proposed by Euler. It is

$$[(3/2)^k] + 2^k - 2$$

in which the square brackets here carry the special meaning that the expression inside them is to be rounded down to the next lower integer. This then, it is proposed, should give the *maximum* number of kth powers necessary to express *all* the integers from one to infinity in the now familiar form. For example, for squares ($k = 2$) the above expression gives

$$[2.25] + 4 - 2 = 2 + 4 - 2 = 4$$

and for the cubes ($k = 3$) gives

$$[3.375] + 8 - 2 = 3 + 8 - 2 = 9$$

reconfirming our earlier findings for these cases. For $k = 4$ through $k = 9$ it gives respectively

$$19, 37, 73, 143, 279, 548 .$$

The only one of these for which there is no independent proof today is, surprisingly, that for $k = 4$. Although it *almost* certainly requires 19 fourth powers to do the job, the absolute proof for this case is still lacking. This is particularly astonishing when we add that an absolute proof has been given for all other k-values up to $k=20{,}000$ and in each case conforms with the formula given by Euler and set out above.

Given all this information it is now interesting to ask whether the 'big gee' phenomenon found for cubes (that is for $k=3$) also exists for larger k values. In other words we want to know whether a smaller number of terms than that given by the Euler equation is sufficient for forming integers which are arbitrarily large. The answer is that the effect undoubtedly does exist for all k greater than 2, and the value of 'big gee' is actually known for $k=4$; it is 16. Thus, although 19 fourth powers are probably necessary to make up all integers, there is some finite number above which 16 will always suffice. Most research on Waring's problem now centers around the calculation of 'big gee' for other values of k. Apart from the squares and the fourth powers no other value of 'big gee' is yet known for certain. In this respect many enormous gaps in our knowledge remain to be closed. For example, for sums of fifth powers ($k=5$) 'big gee' may be as small as 6 or as large as 23, while for seventh powers ($k=7$) it may be any value between 8 and 137. Quite evidently Mr. Waring opened up a veritable can of worms when he first made those few simple speculations all those years ago.

But there is more. For those who have still not seen enough of Waring's problem, the entire question can be re-posed by allowing negative as well as positive terms in the Waring sum. This form was initially termed 'an easier Waring's problem' by its originator (since it *was* easier for the particular case of squares) but it has generally turned out to be anything but that. Except for the squares, where it is known that three terms will always suffice, virtually nothing is known about the 'easier' problem. The proof for the squares follows from a fact noted earlier in this chapter; namely that square numbers are just the sum of consecutive odd integers. It follows that any odd number can always be written as the difference between two consecutive squares. Obviously, therefore, any even number can be expressed by the difference between two squares plus 1 (or plus 1^2 which is the same thing) making three square terms sufficient for the 'easier Waring' problem. For example, those numbers

$$7 = 2^2 + 1^2 + 1^2 + 1^2$$
$$12 = 3^2 + 1^2 + 1^2 + 1^2$$
$$15 = 3^2 + 2^2 + 1^2 + 1^2$$

from our earlier table (for squares) which needed the full 4 positive terms in the original Waring context, are now easily reduced to 3 or less if we allow negative terms as follows:

$$7 = 4^2 - 3^2$$
$$12 = 6^2 - 5^2 + 1^2$$
$$15 = 8^2 - 7^2.$$

For the equivalent problem involving cubes it is thought that 4 will suffice, although this has never been proven. One thing which makes the 'easier' Waring problem more difficult than its regular counterpart is the fact that there are usually so many different ways of arranging terms to make up the same integers if negative terms are allowed. Thus, for example, in the regular problem the integer 2 can be formed from cubes in only one way ($1^3 + 1^3$). In the 'easier' Waring problem at least 23 different ways are known in which 2 can be made up using three cubic terms. Because of these additional complexities the 'easier' problem remains, for the most part, a question for the future.

When one gets to large numbers even the regular Waring solutions can be formed in more than one way. However, one does have to go to surprisingly large numbers before the smallest are found which can be formed from the sum of positive terms in more than one way. We mention this primarily because of a delightful story concerning the brilliant young Indian mathematician Srinivasa Ramanujan who, when he was at one time in hospital, was visited by an English mathematician friend G. H. Hardy who commented that the taxicab which had just delivered him to the hospital was numbered 1729 "a somewhat uninteresting number". Ramanujan immediately responded that 1729 was by no means uninteresting, since it is the smallest integer which can be expressed as the sum of two cubes in two different ways. This is true and the two solutions are

$$1729 = 9^3 + 10^3 = 1^3 + 12^3.$$

As a matter of fact it was Fermat (again) who originally posed the problem of finding more general numbers of this kind. One other example (for cubes) is

$$40,033 = 16^3 + 33^3 = 9^3 + 34^3$$

while the smallest solution in fourth powers is

$$635,318,657 = 133^4 + 134^4 = 59^4 + 158^4.$$

But Fermat, as the reader must by now recognize, was full of these teasers. Other unproved statements which he left to posterity include such gems as $25+2=27$ is the *only* whole number solution of the equation

$$x^2 + 2 = y^3$$

and, similarly, that $4+4=8$ and $121+4=125$ are the only whole number solutions of

$$x^2 + 4 = y^3.$$

Both of these deceptively simple statements are true, but are incredibly difficult to prove. We shall close here with one final Fermat theorem concerning the shapely numbers, again simple to state but as usual not at all easy to prove. It is that no triangular number (other than 1) is also a fourth power. The statement is true and was eventually proved by the French mathematician Adrien Legendre some one hundred and fifty years after Fermat left it for following generations to wrestle with.

13. MAGIC SQUARES AND CUBES

After a chapter on the shapely numbers it seems only logical to consider briefly the age old fascination of assembling the natural numbers 1, 2, 3, 4, ... into patterns such that their sum along various symmetry directions is always the same. The most common such pattern is a square and the literature on the so-called magic squares is already enormous without adding to it. Nevertheless the topic does contains unsolved problems so that in this respect it is still 'alive' and there is still an opportunity for 'pushing back the frontiers'. The extension to three dimensions, leading to a search for magic cubes, is of much more recent origin and is, in many respects, still in its infancy. The three dimensional analogue seems to represent a far more difficult problem and, for many years, not a single example of a magic cube was known. Recently, however, a breakthrough has been made and a new area of number fascination has been opened up for all to enjoy.

A standard magic square is a square array of positive integers starting from 1 and arranged in such a manner that every row and every column (and also the two main diagonals) add up to the same number. If the square has n spaces or 'cells' along each side, then n is termed the *order* of the magic square. Such a square will have a total of n^2 cells so that, in filling them, all the integers from 1 to n^2 will be involved. The 'magic' number which results when any of the rows, columns, or diagonals is added is called the *magic constant* of the square, and it is not difficult to show that this quantity must be just the sum of all the numbers from 1 to n^2, divided by n. A little simple algebra enables this to be reduced to the form $\frac{1}{2} \times (n^3 + n)$.

The square of order 1 is obviously a trivial one consisting of the numeral 1 by itself. The next simplest magic square should be of order 2 and involve the four numbers 1, 2, 3, 4, but a little trial and error will soon convince the reader that no such two-by-two magic square can possibly be arranged. It follows that the smallest and simplest non-trivial magic square must be at least of order three and have a magic constant equal to $\frac{1}{2} \times (3^3 + 3)$ which is 15. A three-by-three magic square can be found but there is only a single example. It is therefore rather special and we give it here:

2	9	4
7	5	3
6	1	8

It can be rearranged by rotations or by reflections (in a mirror perpendicular to the paper), but such processes are not usually considered to generate number sequences which are different from the original in any essential way. The origin of this 'prototype' order-3 magic square can be traced back almost two thousand five hundred years to ancient China and for many centuries it was a mystical Chinese symbol of great significance.

On progressing to squares of order-4 and higher, a true wealth of 'magicness' is revealed and there are no less than 880 different (that is excluding rotations and reflections) magic squares of order-4 alone. So diverse are they that in the sixteenth and seventeenth centuries magic-square-making flourished much as the crossword puzzle does today. The complete list of different magic squares of order-4 was first given in the year 1693. For the record we will present just one of them:

15	10	3	6
4	5	16	9
14	11	2	7
1	8	13	12

The magic constant is 34, as can be seen by inspection, and conforms with our formula $\frac{1}{2} \times (4^3 + 4)$. We have presented this particular square to show that some squares are even more magic than others. For example, not only does this one conform with all the requirements as regards the sum of rows, columns, and diagonals, but over and above this, *any* small two-by-two square of four numbers also adds up to the very same magic constant. It is the search for such additional symmetries which adds delight to the study of magic squares.

Moving on to squares of order-5 there are even greater numbers of magic arrays; so many, in fact, that the total number of different ones (again ignoring reflections and rotations) has only recently been obtained by computer. It turns out that there are an astonishing

$$275,305,224$$

different magic squares of order-5. This is truly incredible and means that the task of classifying them in any meaningful way is monumental. Their magic number is 65 and, being of odd order, they all possess a unique center-cell position. It is of some interest to note which of the 25 integers (from 1 to 25) used in the order-5 problem appears in this special position most often among the complete set of 275,305,224 squares. The answer is the number 13 (which is the 'middle number' half way between 1 and 25) although every one of the 25 allowed integers can appear in the mystical center position and none is particularly rare, although the extreme values of 1 and 25 do appear the least. Interestingly enough there is always an exact balance between 1 and 25, 2 and 24, 3 and 23, and so on in this respect, and this must be so because any magic square of order-5 can be turned into another by subtracting each component number from 26. This is a particular example of a more general rule which is valid for general order-n magic squares, and states that any such square retains its magicness when each component number is changed into n^2+1 minus this number.

One of the more pleasing magic squares of order-5 is

17	14	6	3	25
8	5	22	19	11
24	16	13	10	2
15	7	4	21	18
1	23	20	12	9

This has an extra symmetry associated with it which makes any pair of numbers symmetrically opposite the center add up to 26, or two times the center value itself. In addition, both this and the order-4 square shown above have a particular extra delight. If we tile a floor with them, we can outline any five-by-five square (for the order-5 case) or four-by-four square (for the order-4 equivalent) chosen arbitrarily from anywhere on the complete pattern and it will still be magic.

It is also possible to construct five-by-five magic squares which have smaller three-by-three magic squares contained within them. One such example is

8	22	5	6	24
7	14	15	10	19
25	9	13	17	1
23	16	11	12	3
2	4	21	20	18

which contains, in the innermost nine positions, the smaller magic square

14	15	10
9	13	17
16	11	12

made up from the numbers 9 through 17 inclusive. This smaller square is defined a little differently from the others discussed, since it does not use the integers from 1 up to some limit, but rather a complete sequence of integers starting from a number larger than 1. Since adding a constant number to (or subtracting a constant number from) *every* cell of a magic square cannot possibly alter its magicness, this order-3 square must be just a disguised form of the unique three-by-three magic square set out earlier. Subtracting 8 from each of the cell numbers converts it to

6	7	2
1	5	9
8	3	4

which is just our original order-3 square rotated clockwise by ninety degrees.

For magic squares with order greater than 5 there are seemingly even more different examples than for the order-5 analogues. They are so complex that a complete analysis is not yet available for any order greater than five. It *is* known that there is no upper limit to the size of magic squares so that a largest possible magic square does not exist. This means that the odd looking series

$$1; \quad 0; \quad 1; \quad 880; \quad 275,305,224; \quad ....$$

giving the number of different magic squares of order 1, 2, 3, 4, 5, ... goes on forever, although no more of the terms are yet known. There is no particular difficulty in forming a magic square of arbitrarily large dimension since simple rules exist which allow for the formal construction of a magic square of any size, although the rules generally differ according to whether the array is of odd or even order. As a result there is no special challenge in constructing large magic squares unless one is seeking extra degrees of 'magicness'. As an indication of how simple some of these rules can be let me set out an example which works for all odd order squares.

Start with a numeral 1 in the center position of the bottom row (let us take a seven- by-seven square as an example) and move in a diagonal fashion down and to the right, writing numbers in their natural sequence 1, 2, 3, 4, ... Whenever the next position falls outside the square (as it does for the very first move involving the integer 2) simply transport it across the square to the opposite edge and carry on. For our seven-by-seven example we therefore start as follows:

*	*	*	*	2	*	*
*	*	*	*	*	3	*
*	*	*	*	*	*	4
5	*	*	*	*	*	*
*	6	*	*	*	*	*
*	*	7	*	*	*	*
*	*	*	1	*	*	*

The only remaining rule is that when you can't go because the space you are led to is already occupied by another number (as occurs in the above pattern for the integer 8 which wants to occupy the 1 square) or because you have reached the bottom right hand corner cell and dont know where to go, then simply move up one square and carry on from there with the normal diagonal moves down and to the right. The final seven-by-seven pattern formed by this procedure is

22	31	40	49	2	11	20
21	23	32	41	43	3	12
13	15	24	33	42	44	4
5	14	16	25	34	36	45
46	6	8	17	26	35	37
38	47	7	9	18	27	29
30	39	48	1	10	19	28

Take a moment to follow the pattern through starting from 1 and going all the way to 49. I think you will agree that, looked at in this way, it is extremely simple and, once you memorize the general procedure, you can immediately create odd-order magic squares as large as you wish. Note that these particular squares also have the property cited earlier for which any pair of numbers symmetrically opposite the center add up to twice the center square value. Unfortunately they do not also have the floor tiling property; but how much can you expect from so little effort!

After experimenting awhile with magic squares it is perhaps a natural progression to begin wondering about the possibility of creating a magic cube. As an extension of the magic square concept, a magic cube is a cubic array of positive integers from 1 to n^3 such that *every* straight line of n cells adds up to the same constant value. These straight lines include the space diagonals (between opposite cube corners) and all the face diagonals as well. Although it is by no means obvious whether or not such a cube exists for any value of n (other than $n=1$), it can easily be shown that, if it does, then it must have a magic constant which is the sum of all the numbers from 1 to n^3, divided by n^2. Once again a little elementary algebra enables this to be written in a simpler form, namely $\frac{1}{2}\times(n^4+n)$.

There is, of course, a unique magic cube of order-1, although it is (like its magic square counterpart) completely trivial and without interest - consisting of the number 1 all by itself. Again, as with the magic squares, it is easy to establish that no magic cube of order-2 can exist. For the cubes, however, it is also true that no magic cube of order-3 exists either. The proof is not difficult but we need not concern ourselves with it here. Proceeding to order-4 the problem becomes considerably more difficult, and not until 1972 was it finally established beyond question that magic cubes of order-4 are also non-existent. At this point we are tempted to wonder whether magic cubes exist at all but, happy to relate, they do and the first one was published privately in the year 1905 and was of order-8. It is now believed that perfect cubes probably exist in all orders greater than eight, and perfect cubes have also more recently been constructed of order-7. Whether perfect cubes of order-5 or of order-6 exist, however, still seems to be an open question. None has yet been constructed (as of 1981) although no proof has yet been given that they do not exist. It *is* known that if a perfect cube of order-5 exists then its center cell must be occupied by the number 63.

Most perfect cubes known today are of order-8 and this is almost entirely due to the work of a single investigator, Richard L. Myers. His interest in magic cubes was stimulated at a very early age, and he achieved a great breakthrough for order-8 cubes in 1970 when he was still but sixteen years old and a schoolboy in Pennsylvania. The magic constant for these order-8 cubes is $\frac{1}{2}\times(8^4+8)$ or 2,052, and Myers was able to discover literally millions of them by devising a particular construction procedure. Since so many can be formed, and as each contains no less than 512 separate numbers, we shall not attempt to set any down in detail here. Nevertheless, when you consider that only a handful of examples of magic cubes of any order were known before Myers began his study, the true magnitude of his achievement can be appreciated. Some of Myers' magic cubes have additional symmetries similar to those we cited earlier for magic squares. In some the numbers symmetrically opposite the center always add up to the same total, while others can be separated into smaller component 'sub-cubes' all of which have remarkable properties in common. For example, some can be sliced into sixty four order-2 cubes (of size two-by-two-by-two) with the eight numbers in each of these 'sub-cubes' always adding up to the same constant value. Finally and most significantly these order-8 magic cubes, as composed by Myers, can be used as the basic building blocks with which to assemble perfect magic cubes of order $8^2=64$, and these can again be used to construct even larger perfect cubes of order $8^3=512$, and so on to higher and higher powers of eight. This immediately shows that there can be no largest magic cube of all. In other words magic cubes of arbitrarily large dimension can certainly be constructed.

It is now clear that billions and billions of magic cubes must exist, and it is already difficult to imagine that only a few years ago the known ones were so rare that mathematicians even relaxed the requirements of 'magicness' in order to relieve their frustration and enable them to make some progress by building 'semi-perfect' cubes. In semi-perfect cases one is allowed to ignore the face diagonals in the magic condition and, with this relaxation, semi-perfect cubes of orders as low as 3 do exist. Before the work of Myers, such 'less than perfect' cubes were the objects of much attention. Now, at last, with the realization that enormous numbers of fully perfect cubes are out there waiting to be discovered, the emphasis has begun to move away from the semi-perfect cubes towards the 'real thing'.

14. HOW CAN ANYTHING SO SIMPLE BE SO DIFFICULT?

Before leaving the counting numbers some reference must now be made to what (apart perhaps from Fermat's Last Theorem) is the most famous of all unsolved problems concerning the integers. It is unusually frustrating because this particular hypothesis can be stated in terms which are almost unbelievably simple. First mentioned in an exchange of letters between Euler and the German mathematician Christian Goldbach in the year 1742, it is the suggestion that every even number can be written as the sum of two odd prime numbers. If the integer 1 is not considered a prime (which usually it is not) then the numbers 2 and 4 should be excluded, so that a more correct statement of the conjecture is that all even numbers larger than 4 can be expressed (usually in many different ways) as the sum of two primes.

This statement, which is now usually referred to as 'The Goldbach Conjecture', is obviously not difficult to verify for even numbers which are not too large, and by the year 1900 it had been verified for every even integer up to 10,000. In contrast to this excellent beginning, no essential progress was made on the *general* problem before the 1930's, when it was first established that no more than the sum of 300,000 primes would ever be needed for any finite even number. This was a long way indeed from the Goldbach conjecture but it was at least a general result. Nevertheless, since it can readily be established by simple checking that only two odd primes are required for all the even numbers which can more easily be tested, and (and this is particularly significant) the larger the numbers tested, the more different ways there seems to be of finding two primes with the required property, this first general result appeared to be rather modest progress for nearly two hundred years of effort.

For the very smallest even numbers (greater than 4) the Goldbach conjecture can be demonstrated explicitly as follows:

$$6 = 3 + 3, \quad 8 = 3 + 5,$$
$$10 = 3 + 7, \quad 12 = 5 + 7,$$
$$14 = 3 + 11, \quad 16 = 3 + 13,$$
$$18 = 5 + 13, \quad 20 = 3 + 17.$$

Even for some of these, more than one solution is possible; for example

$$
\begin{aligned}
14 &= 3 + 11 &= 7 + 7 \\
16 &= 3 + 13 &= 5 + 11 \\
18 &= 5 + 13 &= 7 + 11 \\
20 &= 3 + 17 &= 7 + 13
\end{aligned}
$$

For integers only a little larger the number of different representations in terms of two odd primes soon becomes impressively large. A number as small as 48, for example, has five different Goldbach representations in the form

$$48 = 5+43 = 7+41 = 11+37 = 17+31 = 19 + 29.$$

So persuasive is the numerical evidence from modern computer studies that hardly anyone really doubts the validity of the conjecture. But still it has never been proven. For sure the number of primes necessary for which a general proof is available has shrunk very considerably from the 300,000 of the early 1930's, but the fact remains that it has not yet dwindled down to the desired 2. By the year 1937 it had already been established that no more than 67 primes would ever be necessary, and over the years this number was gradually reduced further until the Russian mathematician I. Vinogradoff eventually got it all the way down to four. But the final step from Vinogradoff's four primes to Goldbach's two has still to be taken, and unfortunately it appears to be a mighty stride rather than a simple last step home. Perhaps the fact that Euler could find no way to attack the problem should have warned us that it would be a long long trek to the solution.

Although not helping with regard to the Goldbach conjecture itself, a recent piece of computer investigation seems to suggest that all even numbers which are sufficiently large can even be expressed as the sum of two twin primes. Twin primes, if you remember, are a small subgroup of all the prime numbers, being made up of all primes which have another prime just two integers away. The series of twin primes therefore begins

$$3, 5, 7, 11, 13, 17, 19, 29, 31, 41, 43, \ldots$$

although it is still not known for certain whether this list is infinite. A computer search of all the even integers to one million found that those greater than 4,208 could all be represented by the sum of two primes taken from this restricted group of primes, although there were many exceptions among the smaller numbers. Moreover it was discovered that all even integers larger than 24,098 (and less than

1,000,000) could be expressed as the sum of a pair of twin primes in more than one way, while some could be so decomposed *in more than one thousand ways*. This numerical evidence therefore suggests that for sufficiently large even integers a much more restrictive condition than the Goldbach conjecture may well hold (which would, of course, also imply that the number of twin primes is infinite). And yet the weaker Goldbach conjecture has still not been proven even for numbers which are arbitrarily large.

Interestingly enough, a somewhat weaker theorem relating to the Goldbach problem was established by Vinogradoff in 1937. It is that all sufficiently large odd numbers are the sum of three odd primes. The limiting odd number below which the condition breaks down is not known. As a general proof this is impressive, although it is weaker than the Goldbach theorem since it is very easy to prove it if the Goldbach result is assumed to be true. This follows because we can obviously always subtract an odd prime from an odd number to leave an even number. If this even number is the sum of two odd primes (as Goldbach suggests) then the original number must be the sum of three odd primes. Unfortunately for us the argument does not work backwards, and Vinogradoff's impressive proof for odd numbers still falls short of establishing as true that simple Goldbach statement which was made to Euler back in 1742, and which is overwhelmingly supported by all the numerical evidence available from the fastest of today's computers.

Just how strong the most recent numerical evidence is we shall now consider. It can be judged from a study of what is known as 'The Goldbach Curve' which measures the different number of ways in which the Goldbach condition can be satisfied as a function of increasing even number. Only if this curve ever falls all the way to zero is the Goldbach conjecture refuted. Although the curve is very jagged it does seem to increase in a persistent fashion as we move to higher and higher numbers. In the computer check of even integers up to and beyond 100,000, it is found that *every* even number greater than 4,688 has at least 50 different decompositions, and that every even number above 11,672 has at least 100 of them. In Table 9 below we have listed the even number (which we label N) above which there are *at least* 50, 100, 150, 200, ... etc. different Goldbach decompositions. Although *complete* computer checks have not progressed beyond the values given in Table 9 at the time of writing, some 'spot checks' for larger numbers have been made and are perhaps not without interest. For example, it is known that the number 1,000,000 can be

TABLE 9			
The even number N above which computer searches have established that there are at least n different Goldbach decompositions			
n	N	n	N
50	4,688	400	63,926
100	11,672	450	75,188
150	19,246	500	85,616
200	27,908	550	95,276
250	36,242	600	105,368
300	45,998	650	116,618
350	55,446	700	126,878

decomposed into Goldbach pairs in no less than 5,402 different ways, while the next higher even number, 1,000,002, decomposes in 8,200 different ways, and 1,000,004 in 4,161 ways. From this we can appreciate how jagged the Goldbach curve is on the one hand but, from Table 9, also how its 'low points' increase persistently as we move to larger and larger numbers. In fact a formula has been found which reproduces the Goldbach curve throughout its known range with an average error of only about two to three percent, and a maximum error of only a little more than thirteen percent. Considering the jagged nature of the curve this fitting is rather impressive. The biggest error occurs for the even number 33,038, for which the formula suggests 255 different Goldbach decompositions whereas 'only' 224 are actually found. With all this evidence, the remoteness of the possibility that a large number which has no Goldbach decomposition at all will be found (violating the fomula by a full one hundred percent and completely reversing the steady trend seen in the table) can now be appreciated. However, until the proof comes along, there must always remain the tiniest shadow of doubt and no computer will ever remove it completely.

Just in case you think that the computer evidence is already quite persuasive enough to establish the Goldbach conjecture beyond 'any reasonable doubt', it is perhaps advisable to provide you with a little example (from the Goldbach numerical work itself) which illustrates the danger of 'jumping to general conclusions' from a great wealth of extremely suggestive numerical evidence. In the numerical Goldbach reseach it is found numerically that the even numbers divisible by 6 tend to have more decompositions than the other numbers. One

example has been given above by the number of different
decompositions of the sequence 1,000,000 (with 5,402), 1,000,002
(with 8,200) and 1,000,004 (with 4,161), the middle number being
divisible by 6. It happens that there is a good theoretical reason for the
'general' rule since all prime numbers can be separated into groups of
the form $6n+1$ and $6n-1$ (as we saw in the 'horse racing' section of
Chapter 5). For even numbers divisible by 6, both these groups of
primes can be used for Goldbach decompositions, while only one of the
two groups is 'usable' for other even numbers. On the average,
therefore, there should be about twice as many Goldbach
decompositions for numbers divisible by 6 as for the nearest neighbor
even numbers on either side. The particular example of the series
from 1,000,000 to 1,000,004 seems to bear this speculation out
admirably.

With this as background we find, using the computer, that for even
numbers greater than 36 and all the way up to 80,000 there is not a
single exception to this rule. No number divisible by 6 has less
Goldbach decompositions than either of its nearest even neighbors; not
one in more than thirteen thousand tests. Who then will wager that
this is a general rule? Not many, I suppose, or I would not be telling
the story; and sure enough at 80,082 (which is 6×13,347) comes the
exception. This number has 1,005 different Goldbach decompositions
while its immediate even neighbor on the 'down' side, 80,080, has just
one more at 1,006. Yes, it really does seem essential to have that
watertight general proof before the problem can safely be put to rest; it
seems that numbers are just full of surprises.

Let us gather our thoughts for a moment. What are the facts again?
The Goldbach conjecture is about as simple as one could reasonably
expect any mathematical hypothesis to be. In spite of the warning
given in the previous paragraph, the numerical evidence in favor of its
validity is overwhelming; and yet, in more than two centuries of effort
by the world's best mathematicians it remains unproven. How can any
assertion as simple as this, as elementary in content and as 'almost
certainly' true, be so seemingly impossible to prove? One slender
possibility must still remain, of course, that the proof is so elusive
simply because (in spite of the numerical evidence) the conjecture is
nonetheless false. But are there other possible explanations? Could it
be that the conjecture *is* true but *not possible* to prove? To most people
this suggestion smacks of defeatism, 'sour grapes', or both. Surely if it
is true then it must be *possible* to prove it, even if the proof turns out
to be extremely difficult, complex, or lengthy. Astonishingly enough,

this is not necessarily the case, although only since the 1930's has this fact been recognized.

In order to appreciate the peculiar sort of incompleteness of mathematics which allows theorems to be true but unprovable (although we shall not here be able to give any details of the actual proof of this incompleteness) we must first give a little thought to the manner in which mathematical theory is developed from its very beginning. In what sense is the statement that $3 + 5 = 8$ true? In constructing a logical framework for mathematics it is necessary to start from certain 'self-evident truths' called *axioms*. These cannot be 'proven' since without them we have no frame in which to work or prove anything. The axioms of arithmetic are the definition of unity (that is of 1) and the concept of consecutive integers, coupled with definitions involving the meanings of addition, subtraction, multiplication and division. These axioms are the rules of our 'game' called arithmetic and, once they have been stated, the entire field of arithmetic theory is determined in the sense that it is logically deducible from the axioms. To be entirely precise it would also be necessary to specify the principles of logic which are to be used in the proofs of any propositions but, once this is done, it is then possible to set about looking for proofs of the 'less than self-evident' truths which follow logically within the framework. In this sense we can indeed prove that $3 + 5 = 8$ even if, to most, it would seem to be self-evident enough.

All this seems well and good. What then could possibly go wrong? The first thing to guard against is that there are no logical inconsistencies in the axioms. It must never, for example, be possible both to prove a proposition and its opposite from the 'basic rules', since we should then have demonstrated that the axioms or the logic are not consistent, rendering the whole structure worthless. No such inconsistency is known for the arithmetic structure. But do not be led to believe that checking for logical inconsistencies is a trivial exercise. They can spring up in the most unexpected places.

It has long been realized that certain logical difficulties can lurk just beneath the surface of mathematics, even in association with the counting numbers, and can suddenly jump out to bite the unsuspecting. Consider, for example, the problem of finding the smallest number (that is integer) which cannot be defined in less than twenty words. Since the number of words in the English language (or any other for that matter) is finite, and the number of integers is infinite, there must certainly be numbers out there which take twenty or more words to define. Equally certainly, one would suppose, among this group of integers there must be a smallest. The problem certainly seems to be well-defined and with patience and determination we may therefore solve it. Right? .... Wrong! The number sought, in fact, cannot possibly exist, since its mere definition involves a logical contradiction. For example, if someone claimed to have found it, I could say that if your claim is true and this number really is "the smallest number which cannot be defined in less than twenty words", then the phrase just given in quotation marks defines it, and it contains only twelve words. There is no way out of this dilemma; the mere definition of this peculiar number is self-contradictory. Fortunately, this particular logical contradiction does not arise solely within the bounds of arithmetic, since it involves a combination of language and arithmetic. On the other hand, it does serve as a warning that one must guard closely against the presence of logical inconsistencies in constructing a viable system of arithmetic.

Subtle logical contradictions can arise if great care is not taken when we try to count to infinite numbers using our basic structure. For example, the largest possible number might be thought to count the number of objects in the set of everything; but this set, by its very definition, was shown by the English mathematician and philosopher Bertrand Russell to be just as self-contradictory as the peculiar number discussed in the previous paragraph. On the other hand there do not seem to be any such problems involved with the arithmetic of the finite integers. All appeared to be well for these and, until the year 1931, it was thought that if any question involving the finite integers could be made precise, then the question had an answer. However, in 1931, there appeared a difficult but brilliant paper by Kurt Goedel, a young mathematician at the University of Vienna, which rocked the mathematical world. Goedel showed that some precise questions concerning the finite numbers do not have precise answers within the framework of arithmetic as derived logically from the axioms. More precisely he established that there are propositions which can be expressed purely in arithmetic terms, which can *neither be proven nor disproven* within the existing framework. He further showed that it is impossible to 'patch up the difficulty' in the sense that any 'more complete' or 'improved' set of axioms, constructed specially to remove the problem for one particular set of indeterminate propositions, necessarily leads to the existence of further indeterminate propositions. In a simple phrase, mathematics is, by its very nature, incomplete.

Let us now return to the Goldbach problem. Where do we now stand? Following the work of Goedel it is possible (although it has never been so demonstrated) that the Goldbach assertion falls into the class of 'Goedel incomplete' propositions. By this we mean that it could be true, in the sense that no counter-example to it could ever be found, and yet a general proof may just not exist. On the other hand we must recognise that this is no more than a surmise and that other possibilities exist. For example, the assertion may be false, but the smallest even number which is not the sum of two primes may be so large that it contains more digits than can ever be probed by even the fastest imaginable computer. Or the assertion may be true, but the shortest possible proof may require more pages of mathematics than there are atoms in the universe. In these sort of cases the theorem may be 'theoretically' decidable but 'practically' undecidable. Finally, none of the above situations may hold. 'How can the Goldbach problem be so simple and yet so difficult'; perhaps we are just stupid and one day soon some bright young mathematician will come along and show us all how simple it really is if we just approach it the right way.

15. NEARLY ALL NUMBERS ARE INSANE

In the previous chapter we sketched the manner in which the system of arithmetic for the counting numbers could be built up from a set of simple axioms, or 'self-evident' truths, and developed from these initial concepts in a logical fashion. We must now confront the problem that this set of natural numbers is not always sufficient to enable us to carry out all the operations of arithmetic (like addition, subtraction, multiplication and division) which have been defined. In order to make our number system as useful and self-contained as possible, while still retaining its logical consistency, it therefore becomes necessary to gradually expand the meaning of number itself.

Historically, over the centuries, number systems grew from the simple beginnings of the counting numbers 1, 2, 3, 4, ... first to include the concept of zero, and then to embrace the idea of negative numbers -1, -2, -3, -4, ... The latter are, in some sense, not quite as obviously 'real' as the counting numbers (for example, no-one has seen minus two cows or minus two dogs), but they are useful, since only with them can you *always* perform subtraction. Moreover they can easily be defined and included in the arithmetic framework. Without the negative numbers one could still perform a subtraction like

$$8 - 3 = 5$$

but not one like

$$3 - 8 = ?$$

They are therefore required to make the concept of subtraction a perfectly general one. With them one can write

$$3 - 8 = -5$$

and these negative numbers soon become familiar objects in the abstract mathematical sense and even in the commonplace world of everyday living. The -5 above could, for example, easily be the worldly representation of that 5 dollars which you still need to add to the 3 dollars already in your pocket before that 8 dollar purchase can be made. Historically, however, the step from positive to negative integers was not an easy one to take, and they were not fully incorporated into mathematics until well into the sixteenth century.

With the set of integers now extended to include both positive and negative members (and zero) we can always add and subtract, but we still cannot always divide. Some divisions present no problem, such as

$$6/2 = 3,$$

where we use the 'slash' symbol / to stand for 'divided by'. But what about the arithmetic task of dividing 6 by 7? This problem obviously has no solution at all within the family of positive and negative integers so that, in order to make division perfectly general, it is necessary to enlarge our number system still further. In particular, we need to introduce the concept of *fractions*.

Generally we can solve the problem of dividing any arbitrary integer n (positive or negative) by any other m by simply defining new 'numbers' which we write as n/m. These are the fractions, or more precisely they are the *rational* fractions, implying that they are made up of one *integer* divided by another. An additional convenience of these rational fractions is that they are not new numbers which are completely separate from the integers since they include the latter as special cases. This is easily seen since the fraction n/m is exactly the same number as the integer n if we put m equal to 1. By constructing this new set of numbers we have therefore gained quite a lot but have sacrificed nothing. With them we can always add, subtract, multiply and divide without running into problems. The expanded number system, which now includes all the positive and negative integers and fractions, is called the *rational number system*. Since rational is defined in my dictionary as sane or sensible, these are presumably the sorts of numbers which most people feel happy about. To be truthful, however, *ratio*nal in this context really has nothing whatsoever to do with sanity, and simply implies that all these numbers can be expressed as the *ratio* of one integer to another. One problem does remain and it concerns dividing by zero. This operation cannot be consistently defined within the existing framework since it does not lead to any finite member of the set of rational numbers. To overcome this problem such a division is simply not allowed within the 'rules of the game'.

There are, of course, an infinite number of fractions just as there were an infinite number of integers, but the new rational number system is very different from the old system of natural numbers. For example, if we consider a general fraction n/m in which n is not zero (and m cannot be zero, by definition), then no matter how small this fraction is (say 1/1,000) we can always divide it by 2 (1/2,000) and by 2 again (1/4,000) and again (1/8,000) and so on forever. This means that we can always create a limitless number of fractions between any given small fraction and zero. It follows that the rational numbers must be extremely dense (that is 'close together') in this sense. By defining them we have filled in the gaps between the integers, and it is at first tempting to think that we have filled them in completely. No gap seems to be left between a fraction and its nearest neighbor since, by making the integers n and m in the fraction n/m larger and larger, we can always get as close as we please to any number of interest.

Suppose that we want to get as close to the integer 2 as we can. We could start by trying a fraction like 201/100 and, if that is not close enough for our purpose, follow it by 2,001/1,000 or perhaps 2,000,001/1,000,000. And we can do better and better given the time, paper size, and patience, so that it would appear that (in theory at least) we have the ability to approach 2 just as closely as is possible without actually reaching it. This means, for example, that with the rational numbers we might hope to be able to measure any distance along a piece of string (which mathematicians more formally refer to as a *number line*) with infinite precision just by using the 'sane' numbers. From what we have discussed such a conclusion is almost obvious. It therefore comes as something of a shock to find that (obvious or not) it is not true. There are distances along our piece of string which cannot be measured in rational fractions no matter how large a top and bottom integer we are willing to use. Amazingly, not only are there points along our number line (which is the formal way of saying places where you can cut the piece of string) which cannot be exactly expressed in terms of rational numbers, but there are enormously *more* points of this type than there are points which *can* be expressed in terms of fractions. Alright, I hear you say, give me a 'piece of string' job which cannot be done with fractions.

Suppose that you have four pieces of string, each one foot (or one meter if you are metrically inclined) in length, and you form them into a square to enclose an area of one square foot (or meter). How would you go about cutting four more equal-length pieces to form a square area twice as large? Let us first suppose that the job really can be done with fractions. In this case we make our cuts at a distance n/m (feet or meters) along each of the new pieces of string so that

$$(n/m) \times (n/m) = 2.$$

This is just the equation $n^2/m^2 = 2$ or, rearranging the terms,

$$n^2 = 2m^2.$$

First we note that if the fraction n/m had an even integer at both the top and the bottom, then we could divide each by 2 (over and over again if necessary) until one or the other of n and m (or both) is odd, without affecting the value of the fraction. For example, the fraction 164/64 is exactly the same as 82/32 or 41/16. Let us suppose that this simplification has already been carried out so that n and m are not both even. Looking at the equation $n^2 = 2m^2$ we see immediately that n^2 is 2 times some integer (m^2) so that it *must* be even. Moreover, since only even integers have even squares, n itself must also be even and, therefore, m must be odd. Let us now write the even number n in the form

$$n = 2k$$

thereby defining some new integer k. Substituting $2k$ for n in our defining equation $n^2 = 2m^2$ now gives $(2k)^2 = 2m^2$ which is just $4k^2 = 2m^2$. Dividing both sides of this equation by 2 we finish up with

$$2k^2 = m^2,$$

which tells us that m^2 is 2 times some integer (k^2) and so is even. Once again, since only even numbers can have even squares, it follows that the number m must also be even. But we have earlier determined that m must be odd. We have consequently reached a contradiction. The only logical conclusion is that there just is no rational fraction which will solve our problem and tell us where to cut the string. In other words the rational numbers, as densely spaced as they seem to be, are just not dense enough to measure the length that we require.

The number we are looking for in this string-cutting exercise is the one which, when multiplied by itself, makes 2. Most of us will recognize this number immediately as the square root of 2, and happily write it as $\sqrt{2}$ without bothering too much about the implications of the above difficulties. What we have proved is that $\sqrt{2}$ is a number which can never be expressed as any rational fraction. The proof was first given by Pythagoras and it bothered him greatly. A very similar proof can also be set up for the square roots of 3, 5, 6, 7, 8, 10, ... (any number which is itself not the exact square of an integer) to show that none of these can be expressed as a ratio of two integers either. For those interested, it is given in Appendix 4 at the end of the book.

These are all examples of what we shall consequently refer to as *irrational* numbers. They can never be expressed exactly in the form n/m no matter how large we make n and m. But earlier we stated that we can always get as close as we like to any number by using fractions. This statement remains true; we can get just as close as we wish to $\sqrt{2}$ by using fractions, but the problem is that we can never actually get to it. Thus, for example, the series of fractions 14/10, 141/100, 1414/1000, 14142/10000 sets us out on a fractional trek towards $\sqrt{2}$, and other fractions like

$$14,142,136/10,000,000$$

can be passed along the way but, like a desert mirage, the closer we get to $\sqrt{2}$ the further off it seems to be.

If the rational numbers are in a sense 'sane' numbers, then perhaps we should think of the irrational numbers as 'insane', which was how the Greeks viewed them. How could they possibly deal 'rationally' with numbers which could never be written down (even in principle) either in a decimal base or with any other integer as base? Although they did not realize it at the time they should have been even more worried, since it turns out that there are enormously more irrational numbers on our 'number line' than there are rational ones. This fact cannot be established by a simple count since there are an infinite number of both rationals and irrationals, but we shall show it to be true in our last chapter when we meet head-on the challenge of counting the infinite.

At first sight this over-abundance of irrationals does seem rather surprising since our everyday experience seems to belie their existence. Rational numbers are easy to write down and we use them all the time. Irrational numbers, as far as most of us are concerned in our daily chores of cutting, patching and measuring, just do not seem to exist. We simply avoid them by picking a sane number close enough to do the job. When expressed in decimals, as we shall see below, irrationals go on forever without ever repeating themselves. There just seems to be no way of neatly representing them within our counting system at all. In order to refer to them it becomes necessary to invent special symbols and, in the case of a square root (which is only one of many different types of irrational number), the familiar square root symbol suffices. In other cases, such as the renowned quantity pi (which is the ratio of the circumference to the diameter of a circle, and is also irrational), we merely invent another symbol π. Whenever we are interested in its actual value we are forced to revert to sane numbers, such as fractions or decimals of finite length, and pick one which is close enough for our purpose.

That the mere idea of irrational numbers was worrying to Pythagoras and his early Greek colleagues can certainly be appreciated. Suppose, for example, that a right-angled triangle (so dear to Pythagoras) was formed with the two shorter sides each of unit length. Were they to say that the line which made up the third side (of length $\sqrt{2}$) which they could draw and see right there in front of them, had a length which could never be written down, not even in principle? Surely this did border upon insanity. It is nevertheless a fact and one which, through increasing familiarity, we eventually accept as tolerable, even if it does remain extremely inconvenient. In the following sections and chapters we shall come to accept both the rational and the irrational numbers as friends rather than strangers. They can be highly amusing as well as informative and I trust they will fascinate you with their unexpected, and sometimes quite outrageous, properties.

Let us first think a little about fractions. They are often referred to as being 'proper' or 'improper' according to whether the top number is smaller or larger than the bottom one. Since an improper fraction can always be separated into a whole number part and a proper fraction, only proper fractions (that is fractions with values less than 1) will be of relevance for us. Suppose we divide the bottom number into the top number to obtain the decimal representation of a fraction. Do we notice anything special? Let us pick out a few simple fractions at random and see:

$$
\begin{aligned}
3/8 \ &= \ 0.375 \\
3/7 \ &= \ 0.428571\ 428571\ 428571\ .... \\
2/11 \ &= \ 0.181818181818181818.... \\
5/32 \ &= \ 0.15625
\end{aligned}
$$

It looks as if some of them 'come to an end' while others go on forever, but with the digits eventually *cycling* (that is repeating themselves endlessly in a simple pattern). This is a rather paltry selection from which to draw any convincing conclusions, but the statement is correct. We may have to wait awhile in some fractions before the repetition or cycling begins, and the number of digits in the repeating cycle may sometimes be long but, unless the decimal comes to an end, cycle it will.

How do we know that this is *always* true? No finite number of examples, even if we had the time and patience to work out an impressive list, could ever *prove* this assertion. To do this it is necessary to consider a general fraction, say *n/m*. By dividing *m* into *n* in the usual 'long-division' fashion we can turn *n/m* into its decimal form. At each step of the dividing process different 'remainders' are found, and all these remainders must naturally be smaller than the number (*m*) which is doing the dividing. Only two possibilities arise. Either a remainder of zero appears at some point or it doesn't. If it does, the dividing process terminates and the decimal 'comes to an end'. If it doesn't, then the remainders are numbers between 1 and m−1 and appear in some particular order. Therefore, in m−1 steps or less, a remainder must turn up which has occurred before. When this happens the entire dividing procedure repeats itself, and continues to repeat itself endlessly; in other words cycling begins. The maximum number of digits in the repeat cycle is therefore seen to be one less than the bottom number (or *denominator*) of the fraction. Not all fractions with the same denominator *m* necessarily have this longest possible repeat cycle. Thus, with *m* = 3 for example,

$$1/3 = 0.333333333...$$

$$2/3 = 0.666666666...$$

and the longest possible repeat cycle of 2 never arises. It turns out that *m* = 7 is the smallest denominator for which the maximum repeat distance occurs, and one example for this case has already appeared in the decimal expansion of 3/7 given above, which has the repeat sequence 428571 with six digits.

Thus, all fractions when expressed as decimals either come to an end or cycle. Is the reverse statement also true? Do all finite-length and cycling decimals represent rational numbers (that is fractions)? For terminating decimals the answer is obviously yes, since we only need to divide the digits by the appropriate power of ten to get the required fraction; for example,

$$0.314826 = 314,826/1,000,000.$$

We can also show that the answer for cycling decimals is again yes. We shall illustrate the method of deducing the relevant fraction for this case by using a particular example, but it is apparent that the method is quite general and can be carried out for any cycling decimal whatsoever. Consider the cycling decimal

$$c = 0.7911911911911.....$$

Multiplying it in turn by 10,000 and by 10, we get

$$10{,}000c \quad = \quad 7911.911911911...$$

$$10c \quad = \quad 7.911911911...$$

two numbers with exactly the same (infinite length) decimal part. Subtracting the bottom equation from the top removes the infinite decimal and gives

$$9{,}990c = 7{,}904$$

from which we see immediately that the cycling decimal number c is just the fraction 7,904/9,990. Another one of some aesthetic interest might be

$$c = 0.9876543210\ 9876543210\ ......$$

By exactly the same method we find two multiples of this number with the same decimal part, namely

$$10{,}000{,}000{,}000c \quad = \quad 9876543210.98765432109876543210.....$$

$$c \quad = \quad 0.98765432109876543210.....$$

Once again subtracting the bottom equation from the top produces a much simpler equation with the endless decimal removed; in this case

$$9{,}999{,}999{,}999c = 9876543210$$

from which we find the fractional form

$$c = 9{,}876{,}543{,}210/9{,}999{,}999{,}999.$$

Since all finite-length and cycling decimals are rational numbers and conversely all rational numbers are terminating or cycling decimals, it must now follow that the irrational numbers (like the square root of 2) have a decimal form which continues forever without cycling. The irrationals therefore can not be expressed using a finite number of digits either by fractions or by decimals. With this little demonstration we begin to realize why there must be so many more irrational numbers than rational ones. The rational numbers all have a very special decimal property. They either terminate (which is really only another way of saying that they cycle zeroes forever) or they cycle in the more conventional fashion. It is not difficult to appreciate that there are countless different ways of interfering with any particular cycling decimal to ruin the cycling property. Once the cycling has been disrupted by any means then we have generated, by definition, an irrational number. Thus, from the decimal point of view, the rational numbers are a very special and tiny subset of all the possible decimal numbers. In the 'tongue in the cheek' language of this chapter heading, nearly all numbers are insane!

16. CYCLIC NUMBERS AND THEIR SECRET

The number 142,857 has long been recognized by those interested in number oddities as one of the most remarkable of integers. When multiplied by any number from 1 to 6 the result always consists of the same digits and, more remarkably still, these digits always appear in the same cyclic order, but with each number commencing at a different point. Thus we find

$$
\begin{array}{rcl}
1 \times 142,857 & = & 142,857 \\
2 \times 142,857 & = & 285,714 \\
3 \times 142,857 & = & 428,571 \\
4 \times 142,857 & = & 571,428 \\
5 \times 142,857 & = & 714,285 \\
6 \times 142,857 & = & 857,142
\end{array}
$$

Because of this rather striking property 142,857 is called a *cyclic number,* and it is interesting to proceed further to multiply it by 7, recognizing that there just are no more cyclic combinations of the six digits left to appear. On checking it out we obtain a very different, but still striking, result namely

$$7 \times 142,857 = 999,999.$$

Is this phenomenon just an isolated numerical freak, or is there a deeper significance to this almost supernatural pattern? Well, there is at least a clue in the final relationship obtained above involving the multiplication by 7. The resulting string of nines is telling us that the cyclic number 142,857 is equal to 999,999/7, or very nearly $10^6/7$. This is only another way of saying that the decimal representation of 1/7 is very nearly 0.142857.

We saw in the last chapter that all rational fractions like 1/7, when expressed as decimals, either terminate at a particular decimal position (like $1/4 = 0.25$) or they go on repeating themselves endlessly in a cycling pattern. The number 1/7 is in the latter category and its exact decimal expansion begins

$$1/7 = 0.142857\ 142857\ 142857\ 142857\ ....$$

with the dots implying a repeating pattern going on forever. We note that the number of digits in the repeating part of the decimal pattern is one less than the denominator of the fraction and, as we also saw in the last chapter, there is a general rule which says that 'one less than the denominator is the largest number of repeating digits which any fraction can possibly have'. Furthermore, it can be shown that only fractions which have a prime number for a denominator are possible candidates for this maximum repeat distance or *period,* as it is called, although not all the prime numbers actually have this property. Looking at the smallest prime number denominators we find that $1/2 = 0.5$, $1/3 = 0.33333$ ...., and $1/5 = 0.2$, so that 7 is the smallest prime denominator which exhibits the maximum cycling period.

In the light of this preamble it seems fairly clear that the unusual cyclic-number property of 142,857 must be closely related to the decimal representations of the fractions $1/7, 2/7, 3/7, ..., 6/7$. In detail these representations are

$$1/7 = 0.142857\ 142857\ 142857\ ....$$
$$2/7 = 0.285714\ 285714\ 285714\ ....$$
$$3/7 = 0.428571\ 428571\ 428571\ ....$$
$$4/7 = 0.571428\ 571428\ 571428\ ....$$
$$5/7 = 0.714285\ 714285\ 714285\ ....$$
$$6/7 = 0.857142\ 857142\ 857142\ ....$$

and it does not take a large leap in the imagination to suspect that the next larger prime number p which has a decimal representation of $1/p$ with the maximum periodicity of $p-1$ might be a good candidate for exhibiting the cyclic-number property on an even grander scale. Unfortunately there does not seem to be any simple rule dictating which prime numbers p will have this property and which will not, so that one just has to check each prime out by long division to see. It turns out that the next higher prime number with the desired property is 17, and the 16-cycle repeat decimal pattern of $1/17$ is

$$1/17 = 0.0588235294117647\ 0588235294117647\ ....$$

A little multiplication now reveals that the magical cyclic-number property is indeed again valid for the 16 cycling digits of this infinite decimal. But now the mysterious property is true when multiplying by all the integers between 1 and 16 inclusive. The multiplication chart begins with

$$1 \times 0,588,235,294,117,647 = 0,588,235,294,117,647$$
$$2 \times 0,588,235,294,117,647 = 1,176,470,588,235,294$$
$$3 \times 0,588,235,294,117,647 = 1,764,705,882,352,941$$
$$4 \times 0,588,235,294,117,647 = 2,352,941,176,470,588$$

and ends with

$$16 \times 0,588,235,294,117,647 = 9,411,764,705,882,352.$$

We are now no longer surprised to find that

$$17 \times 0,588,235,294,117,647 = 9,999,999,999,999,999.$$

A search for still larger prime numbers with the same 'magic' cyclic properties reveals that they are not at all rare. In fact no less than seven more prime numbers smaller than 100 generate cyclic-numbers in the manner shown for 7 and for 17. They are 19, 23, 29, 47, 59, 61, and 97. On the average about one prime number in three seems to generate a fraction $1/p$ with the desired maximum period $p-1$, but there seems to be some doubt as to whether these 'cyclic primes' get denser or thin out among the primes in general as one progresses to larger and larger values. With this insight we can now learn exactly why the cyclic-number properties *must* occur in fractions like 1/7, 2/7, 3/7, ... etc. If we set out the 'long division' of the fraction 1/7 we see that the remainder at each step can only be one of the numbers between 1 and 6. What is more, as soon as any remainder is repeated, the entire sequence of remainders must begin again. Now in the fraction 1/7 all the possible remainders 1, 2, 3, 4, 5, 6, actually do occur (in the sequence 3, 2, 6, 4, 5, 1) before the first of them (3) is repeated, so that this six-remainder sequence will repeat endlessly to produce the associated cycling decimal sequence 1, 4, 2, 8, 5, 7. Now consider the fraction 2/7. Setting out the long division again we find that it undergoes exactly the same cycling process as was found for 1/7, but starting at a different place along the sequence. As a result it produces the same cycling decimal digits in the same order, but with the cycling part starting at a different place along the sequence. For all the fractions between 1/7 and 6/7 you must therefore get exactly the same order of digits in the remainder sequence (and therefore also in the decimal itself) but with the cycle starting at a different place each time. It is now clear why the 'magic' of cyclic numbers must work and, using the special sequence of prime denominators (7, 17, 19, 23, 29, ... etc.) which exhibits a maximum repeat cycle, we can now generate cyclic number multiplication tables as long and as impressive as we wish.

The cyclic numbers have many other surprising and intriguing properties and we shall now touch upon a few more of them. One we have already noticed in connection with the first cyclic number 142,857; it is that when multiplied by its 'generating prime' (in this case 7) it just produces a lengthy string of nines. More interesting is the fact that all the cyclic numbers have an even number of digits and, when split in half, give two numbers which always add up to another row of nines. For example, using the 1/17 cyclic number 0,588,235,294,117,647 we find

$$
\begin{array}{r}
05882352 \\
\underline{94117647} \\
99999999.
\end{array}
$$

This result has its usefulness because, if you are doing the long division to obtain a lengthy cycle (say 1/19), you already know how many digits it has in it (in this case 18). You can therefore stop the long division at the half-way point and, by looking at those digits already obtained, continue by writing down their differences from 9. Thus, for 1/19 we find the decimal representation

$$0.052631578...$$

the hard way, and finish the job

$$0.052631578\ 947368421\ ...$$

in a flourish.

Another strange property of the cyclic numbers is that each can be generated *in many ways* by adding together a set of numbers written diagonally in a geometric progression (which means that each number is some constant times the one before it). For example, starting with the number 14, doubling at each step, and writing each number so that it 'moves over' two digits to the right, we produce a strange addition sum

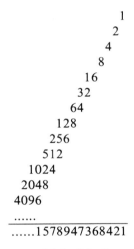

```
    14
    28
     56
     112
      224
       448
        896
         1792
          ......
   _____
   142857142857......
```

which generates the smallest repeat-distance cyclic number. As another example, starting from 1, doubling at each step, coming down diagonally from the top right and moving over only one place to the left for each new line, we generate

```
                1
               2
              4
             8
           16
          32
         64
       128
      256
     512
   1024
  2048
 4096
   ......
   _____
   ......1578947368421
```

another curious addition sum which this time gives the repeating cyclic sequence for 1/19 which was set out in full a few lines above.

But even now we have seen only the 'tip of the iceberg' of fascination which the cyclic-numbers generate. To this point we have been discussing only the true cyclics, more properly called the *cyclic-numbers of order one*. In addition to these there are whole families of other related cyclics which each have similar remarkable properties. The next simplest group of cyclics is the group of order two. The smallest of these is generated by the fraction 1/13, the decimal form of which is

$$1/13 = 0.076923\ 076923\ 076923\ ....\ .$$

Instead of having a repeat distance of 12, as a cyclic of order one would have, it repeats twice as often (that is with period 6). If we multiply this cycling six-digit number 076923 by the numbers 1 through 12 we find a most interesting pattern as follows:

1×076923	=	076923	2×076923	=	153846
3×076923	=	230769	5×076923	=	384615
4×076923	=	307692	6×076923	=	461538
9×076923	=	692307	7×076923	=	538461
10×076923	=	769230	8×076923	=	615384
12×076923	=	923076	11×076923	=	846153

Half the products (specifically those in the left column) are the six cyclic forms of the original number 076923, while the other half are the six cyclic forms of another equal-length number 153846. We note that each of these two numbers can be split in half and added to get 999, and we are again no longer surprised to discover that

$$13×076923 = 999999.$$

All these properties are quite general for cyclic numbers of order two. And how do we find more second order cyclics? They are also generated by a subset of prime fractions $1/p$. In addition to the prime $p=13$, the other prime numbers smaller than 100 which generate second order cyclics are 31, 43, 67, 71, 83, and 89.

In the same manner one can now go on to define, and find, cyclic numbers of orders three, four, five, and even higher. The smallest prime number producing a third order cyclic is 103. The fraction 1/103 cycles with a period of 34 digits (instead of the 102 digits a first order cyclic would have). When this 34-digit number is multiplied in turn by all the integers between 1 and 102 the results fall into three sets, each containing 34 cyclic forms of a 34-digit number. The smallest primes generating cycles of order four through ten are respectively 53, 11, 79, 211, 41, 73, and 281. All the prime numbers less than 100 have now been covered with the exception of $p=37$; its cycling decimal 0.027027027... has a period of only 3, and 37 is therefore a cyclic number of order 12. Of the higher order cyclics by far the easiest to probe is the fifth order one generated by 11. The decimal representation of 1/11 is

$$1/11 = 0.09\ 09\ 09\ 09\ 09\ 09\ .....$$

with a period of 2, or one fifth of the 10-digit cycle which would make it a first order cyclic. Multiplying 09 in turn by the integers from 1 to 10 produces the five pairs

$$(09, 90);\ (18, 81);\ (27, 72);\ (36, 63);\ (45, 54).$$

Once again, each of the numbers (09,18,27,36,45) involved in the cycles can be separated into two halves which add together to give nothing but nines. I now leave it to the interested reader to go ahead and have fun with cycles of other orders and cycling periods.

On the other hand I do not wish to imply that a number has to be a prime in order to generate intriguing cycles. The decimal forms of quite arbitrary fractions nearly always have some unexpected fascination in their cycling behavior. Consider, for example, the fraction

$$1/21 = 0.047619\ 047619\ 047619\ .....\ \ .$$

It hasn't been chosen for any special reason other than the fact that its cycle period is of a manageable size, neither trivially short nor wearyingly long. If we add the cycling digits together one at a time until they repeat we get

$$0 + 4 + 7 + 6 + 1 + 9 = 27.$$

Let us now do this again, but taking the digits two at a time, then three, then four, then five, continuing along the endless decimal in each case until the numbers start to repeat. We find for the two digit case

$$04 + 76 + 19 = 99,$$
$$47 + 61 + 90 = 198,$$

for the three digit case

$$047 + 619 = 666,$$
$$476 + 190 = 666,$$
$$761 + 904 = 1,665,$$

for the four digit case

$$0,476 + 1,904 + 7,619 = 9,999,$$
$$4,761 + 9,047 + 6,190 = 19,998,$$

and finally for the five digit case

$$04,761 + 90,476 + 19,047 + 61,904 + 76,190 + 47,619 = 299,997.$$

All these sums are intimately connected with the number 9. For example, they are all exactly divisible by nine, and many contain far more nines than one might reasonably expect to occur by pure chance. In fact, 14 of the 32 digits involved in the sums (or over 40%) are nines.

A coincidence you say. Well, after 21 the next non-prime integer to have a middle-sized repeat distance is 26. In decimal form 1/26 is

$$0.0384615\ 384615\ 384615\ ....$$

and the results of adding the cycling digits together one, two, three, and more at a time are shown in Table 10.

TABLE 10
1/26 = 0.0384615 384615 384615 ......
$3 + 8 + 4 + 6 + 1 + 5 = 27$
$38 + 46 + 15 = 99,\quad 84 + 61 + 53 = 198$
$384 + 615 = 999,\quad 846 + 153 = 999,\quad 461 + 538 = 999$
$3,846 + 1,538 + 4,615 = 9,999,\quad 8,461 + 5,384 + 6,153 = 19,998$
$38,461 + 53,846 + 15,384 + 61,538 + 46,153 + 84,615 = 299,997$

The findings here are even more striking than they were for 1/21. Once again each sum is an exact multiple of nine and now no less than 23 of the 31 digits contained in the various sums are nines (this is close to three quarters).

These curious properties of the rational numbers in their decimal forms persist even if we move on to numbers for which the repeat cycle is much larger. Up to 50 the longest repeat cycle for a non-prime integer is that for 1/49, which has no less than 42 digits in its cycle. This numerical beast is shown in Table 11 where we have also recorded its sums of cycled digits taken one, two, three, ... etc. at a time.

TABLE 11
1/49 = 0.020408163265306122448979591836734693877551 020408.....

$$0 + 2 + 0 + ... + 5 + 5 + 1 = 189$$

$$02 + 04 + ... + 75 + 51 = 990, \quad 20 + 40 + ... + 55 + 10 = 1,089$$

$$020 + 408 + ... + 877 + 551 = 6,993$$
$$204 + 081 + ... + 775 + 510 = 6,993$$
$$040 + 816 + ... + 755 + 102 = 6,993$$

$$0,204 + 0,816 + ... + 8,775 + 5,102 + ... + 9,387 + 7,551 = 99,990$$
$$2,040 + 8,163 + ... + 7,755 + 1,020 + ... + 3,877 + 5,510 = 109,989$$

$$02,040 + 81,632 + ... + 46,938 + 77,551 = 2,099,979$$

Again we see that the sums are all exactly divisible by 9 and that (at least as far as we have gone, which is to summing the digits five at a time) the numeral 9 appears in the sums in about 50% of the available digit positions.

But the mysteries are even deeper than we have yet realized. Consider, for example, the result in Table 11 for summing the 1/49 cycle in groups of five. It is

$$2,099,979.$$

If we keep only the number of digits on the right up to five (the 'grouping' number); that is

$$99,979,$$

'carry over' the unused digits 20 to the appropriate columns on the right, and add them as follows;

$$99,979$$
$$+ \quad 20$$
$$99,999$$

we get *all* nines. Try this again for the group of five sum in the cycles for 1/21 and 1/26. They both give

$$99,997$$
$$+ \quad 2$$
$$99,999.$$

Now move down to the cycles taken in groups of four, and keep only the rightmost four digits of the relevant sums. Once again 'carry over' any unused digits from the front to the columns at the right and add. For the 1/49 cycle we get the two results

$$9,990$$
$$+ \quad 9$$
$$9,999$$

and

$$9,989$$
$$+ \quad 10$$
$$9,999$$

while for the fractions 1/26 and 1/21 the corresponding two results are

$$9,999$$
$$+ \quad 0$$
$$9,999$$

and

$$9,998$$
$$+ \quad 1$$
$$9,999$$

in both cases.

   This procedure obviously has a great tendency to produce a whole string of nines. Now the truth is that it does not *always* do so, but does so often enough to impress the mathematicians. Moreover one need not take only simple reciprocals of integers (that is *one* over some number n) in order to obtain these effects, any rational number m/n will do, just as long as the decimal forms cycle with other than a trivially short period. The full significance of these patterns is still not understood although the repeated appearance of nine, rather than any other number, is known to relate to its being one less than the base (ten) to which our familiar counting system operates. If we count to the base 8, for example, using only the eight numerals 1,2,3,4,5,6,7,0, then it is 7 which appears in the anomalous role. But why the intriguing pattern of nines occurs so frequently (but not always) in the operations with decimal fractions, and what it signifies about our system of rational numbers, is still rather a puzzle, even though the first publication on the phenomenon appeared way back in the year 1802.

17. PI, A TRANSCENDENTAL NUMBER

Perhaps the most familiar of all the irrational numbers is the number π, which measures the ratio of the circumference of a circle to its diameter. Since it is irrational (although this fact has not always been known) it cannot be expressed by using a finite number of digits either as a fraction or a decimal. We therefore denote it by a special symbol for convenience. This symbol π, pronounced 'pi', was not actually used until the eighteenth century (in fact it was first written by Euler) although interest in the quantity which it represents, and in the question concerning its possible irrationality, goes back much further.

The first record of any interest in the ratio of the circumference to the diameter of a circle comes from Egyptian and Babylonian times, some four thousand years ago. By a direct physical measurement on specimen circles it was apparent that the value was fairly close to the fraction 22/7, and this is the numerical representation most often met in antiquity. In decimal form 22/7 is

$$3.142857\ 142857\ 142857\ ....$$

and is only about 0.04% larger than the true value, which is now known to have a decimal form which begins

$$\pi = 3.141592653589.......\ .$$

It follows that the simple fractional approximation 22/7 is quite adequate for most everyday purposes.

This practical method of measurement, however, does not allow us to zero-in on more accurate decimal representations for pi in any well-defined manner. The first step towards this end was taken by the ancient Greeks (the actual gentleman was Archimedes of Syracuse) and the basic idea was that regular polygons (that is squares, pentagons, hexagons etc. with 4, 5, 6, ... equal sides and equal angles) could be constructed both inside and outside a given circle to touch it in a symmetric fashion. By trigonometric methods it was then known how to calculate the perimeters of these polygons no matter how many sided they were, with limits set only by the time, patience, and fortitude of the mathematician doing the calculation. Now the polygons inside the circle must obviously always have perimeters less than the circumference of the circle, while those outside must always have perimeters greater than this same circumference. Since both of these perimeters gradually approach the value of the required circumference as the number of sides in the polygons increases, then the true value of pi could be bracketed between two calculated numbers which were coming ever closer together. Archimedes himself worked with polygons up to 96 sides and, in this manner, was able to show that pi was actually less than the common fractional value 22/7 but greater than 223/71. The average value of these two fractions, which is the best estimate of pi available by the method using 96-sided polygons, is

$$\pi = 3.1418511....$$

and is greater than the true value of pi by only 0.0082 percent.

Once a method such as the above has been established, it is purely a matter of computational ability and perseverence which determines the number of decimal places to which pi can be calculated. Today this translates to computer programming know-how and a matter of cost. If other methods are found which are more efficient than the polygon technique so much the better. But some of you may already be asking why one needs to go on calculating more and more decimal places once enough significant figures are known for all practical requirements. One can always give the 'because it is there' rationalization of mountaineers, but in earlier centuries at least there was an additional incentive. We have already stated that pi is an irrational number, but this fact is not obvious and is by no means simple to demonstrate. After Pythagoras had established the existence of irrational numbers, it was natural to ask whether pi was one, or whether it was really an (as yet unknown) rational fraction. If the latter were true, then there was a specific goal in continuing the calculations; maybe one could calculate pi *exactly*. The question therefore became 'will the decimal form for pi ever begin to cycle?' If it does then pi is, in all probability, a rational number whose exact form as a fraction can be established by the method discussed in Chapter 15.

The incentive of possibly being the first to find the *exact* value of pi was enhanced over the years by the appearance of several improved methods for calculating the decimal sequence. By the second century A.D. the great astronomer Ptolemy was using the approximate fractional representation

$$377/120 = 3.141666... ,$$

accurate to better than one part in forty thousand, while there is evidence from fifth century China that a value correct to six decimal places had already been obtained there. Nothing better than this was attained in Europe until well into the sixteenth century. It was then that the fractional representation

$$355/113 = 3.14159292...$$

appeared. This is about the closest one can come to the true value of pi with a really simple fraction. It is accurate to six decimal places, being too large by about one part in twelve million. Accuracy to no less than ten decimal places is provided by the somewhat less simple fraction $312689/99532$.

Something of a milestone in the history of pi was achieved in the sixteenth century by the French mathematician Francois Vieta. Although his ideas were still rooted in the concept of the Archimedean polygons (in fact he was one of the last mathematicians to use this method for furthering the theory of pi) he succeeded in relating the area of an n-sided regular polygon to that of a $2n$-sided one. As a result he was able to represent the number of polymer sides going to infinity by an endless sequence of mathematical operations. At infinity the limiting polygon *is* a circle and the limiting area is therefore related exactly to the number pi. In this way Vieta was able to write an expression for pi which was exact in principle, but which involved an endless sequence of mathematical operations (actually multiplications). The specific expression is quite complicated and we shall not give it in detail, but its importance was that it established that pi could be *exactly* related to an infinite sequence of numbers multiplied together.

In the following century, with the invention of the calculus (independently by Sir Isaac Newton and Baron Gottfried von Leibnitz), a whole multitude of infinite sequence relationships for pi were produced. Some were extremely simple but approached their limiting values (mathematicians use the word *converged)* very slowly. Others were more complicated, but converged sufficiently quickly to enable the search for a possible cycling in pi to be investigated in earnest. Of the simpler examples we may quote one involving endless multiplications, namely

$$\frac{\pi}{2} = \frac{2\times2\times4\times4\times6\times6\times\cdots}{1\times3\times3\times5\times5\times7\times\cdots}$$

which dates from the year 1655, and one involving additions,

$$\pi/4 = 1 - 1/3 + 1/5 - 1/7 + 1/9 - \cdots$$

In each case the dots imply that the sequences are to be continued forever in the manner established by the obvious pattern of the first few terms. The latter sequence is perhaps the simplest of all the infinite sequences for pi and it was, in fact, the very first *series* (that is expression involving *addition and subtraction* of terms rather than multiplication and division) to be found; it appeared in the year 1671. Unfortunately it was virtually useless for the digit hunters because it converges extremely slowly, as you will appreciate if you calculate a few terms from it yourself. A much better series from the point of view of quick convergence, which appeared only shortly after its more elegantly simple predecessor, is

$$\pi/6 = x + (1/2)(x^3/3) + (1/2)(3/4)(x^5/5) + (1/2)(3/4)(5/6)(x^7/7) +$$

when the value of x is put equal to $1/2$. It is attributed to Sir Isaac Newton, although Newton himself never considered digit hunting to be worthy of his interest. However others differed and, with the help of series of this kind, the growth of the known decimal expansion for pi can be documented as going from the nine or so decimal places known to Vieta in 1580, to 35 decimals in 1620, 72 in 1700, 100 in 1706, and 127 in 1717. And all this activity was primarily carried out in an effort to find a cycling of the digits, one which never came but which, if it had, would have meant fame indeed for the discoverer.

Finally, in the year 1761, the search for digit cycling in pi had to be abandoned when the German mathematician Johann Lambert succeeded in proving that pi could not possibly be a rational number. His proof was not simple and cannot be sketched in a book of this nature. But an unassailable proof it was and this should have ended any concern about the decimal form of pi beyond the 127th decimal place, but it did not. There are always those who have the mountaineer's philosophy, and the quest for extending the known decimal places of pi continued even though, as Isaac Asimov has pointed out, you could measure the circumference of the known universe (at least in theory) to the accuracy of a millionth of an inch using only the first thirty five decimal places of pi which were already known in the year 1620. By 1873 the expansion of pi had been continued (by hand!) to an almost unbelievable 700 decimal places; a record which perhaps not surprisingly stood until the dawn of the computer age. Then, in the 1950's and later, the number of known significant (or perhaps we should better term them insignificant) figures exploded into the thousands, tens of thousands, hundreds of thousands, and finally past a million. The millionth decimal place was first achieved in 1974 using about twenty four hours of computer time; it is, by the way, a 1. But how can all of this effort possibly be justified today? Well, if justification is needed, the major drive behind the more recent extensions has been an interest in the digit patterns and numeral distributions which occur in irrational numbers. We shall pursue these further in the following chapter, but let us first go back again to the eighteenth and nineteenth centuries to see that a continued interest in pi after 1761 was not restricted only to that small group of persistent digit hunters. To backtrack even further, perhaps the most famous series of all for pi (or more precisely for the square of pi) is that deduced by Euler in 1736. It is

$$\pi^2/6 = 1 + 1/1^2 + 1/2^2 + 1/3^2 + 1/4^2 + 1/5^2 + \cdots$$

and has both a simplistic beauty and a tolerably quick convergence. In this respect it is interesting to note that the fact that pi is irrational does not necessarily mean that the square of pi is also irrational; after all $\sqrt{2}$ is irrational but $(\sqrt{2})^2 = 2$ is certainly not. Thus, some justification of a search for cycling using this Euler series could still have been made after 1761 (I don't know whether it was!) but it is now known that π^2 is also irrational as is *any* power of pi involving a natural number 1, 2, 3, 4, 5, ...

One very famous problem connected with pi which the proof of its irrationality did not settle was the old Euclidean favorite of 'squaring the circle'. The problem, as first posed by Euclid, was this. Construct a square which has exactly the same area as a given circle by using only a straight edge and compasses. By the year 1761 it was generally believed that the task could not be performed in a finite number of steps, but no-one had actually proved it. What has the nature of pi got to do with this problem you may ask? Well, in order to carry out the construction of the desired square for a circle of unit radius, it is necessary to construct a line of length L the square of which is equal to the area of the circle. This area is just pi times the radius squared (from simple geometry) or, for our case of unit radius, just pi. We therefore need to construct a length for which $L^2 = \pi$, or equivalently $L = \sqrt{\pi}$. Now all line segments which can be constructed from a given unit line by a finite number of straight edge and compass operations have lengths which are the solutions of equations of the general form

$$a_0 + a_1x + a_2x^2 + \cdots + a_nx^n = 0,$$

where the coefficients $a_0, a_1, ..., a_n$ are integers. These sorts of equations are called *algebraic* equations. We now ask what kinds of numbers can appear as solutions of algebraic equations? Well, equations like

(1) $\qquad 2 - x = 0,$
(2) $\qquad 9 - x^2 = 0,$
(3) $\qquad 2 + 3x + x^2 = 0,$
(4) $\qquad 1 - 5x + 6x^2 = 0,$

are all examples of this type and these have solutions in integers ($x = 2$ in equation (1), $x = 3$ or $x = -3$ in equation (2), and $x = -1$ or $x = -2$ in equation (3)) or in fractions ($x = 1/2$ or $x = 1/3$ in equation (4)). But so also are equations like

(5) $\qquad 2 + x^2 = 0,$
(6) $\qquad 2 - x^2 = 0.$

and, while equation (5) has no solution at all in terms of either rationals or irrationals, equation (6) has irrational number solutions $x = \sqrt{2}$ and $x = -\sqrt{2}$. Evidently solutions of algebraic equations can be both rational and irrational. But can *all* rational and irrational numbers appear as solutions of this type of equation? More specifically, can $x = \pi$ or $x = \sqrt{\pi}$ be the solution of such an equation? This indeed is the important question because if they cannot then, from what we have already said, squaring the circle is not possible.

The sorts of numbers which can appear as solutions of algebraic equations are, not surprisingly, called *algebraic numbers*. It can be shown without much difficulty that all rational numbers can certainly appear in this way. In other words, *all* rational numbers are algebraic numbers. Whether there are some types of irrational number which can never appear as solutions of algebraic equations is a question for which, in the eighteenth century, the answer was just not known. However, it had to be considered a possibility, and eventually the term *transcendental number* began to appear to denote possible numbers which might not be algebraic. In this new language the question of whether the circle could be squared might therefore be cast in the form 'Is pi a transcendental number?'

At the end of the eighteenth century and into the nineteenth century the very concept of a transcendental number was new. The idea that some numbers might be even more beyond reason than the older generation of irrationals like $\sqrt{2}$ and $\sqrt{3}$ was already suspected in Euler's time, but it was not a comfortable notion. In fact it was not obvious that they existed at all. It was certainly not difficult to write down *equations* which were not algebraic. Simple trigonometric ones involving sines and cosines such as

$$\sin(x) = 1/2$$
or
$$\cos(x) = 1/2$$

for example. Since neither $\sin(x)$ nor $\cos(x)$ can be expressed as an algebraic series in x in a finite number of terms (a fact first demonstrated by Newton in a famous treatise in 1665) these equations are certainly 'transcendental' if by that term we mean 'not algebraic'. So also are logarithmic equations like

$$\ln(x) = 2$$

for the same kind of reason. But do the solutions of transcendental equations have to be transcendental numbers? Obviously not always, since it is well known that the sine of zero is zero, as also is the logarithm of 1. This means that the equations $\sin(x) = 0$ and $\ln(x) = 1$ both have the solution $x = 0$, and there is certainly nothing frightening or transcendental about the number zero; it is the solution of countless algebraic equations as well, such as

$$x^2 = 0$$
and
$$x + x^3 = 0.$$

Obviously, the fact that transcendental equations existed did not necessarily imply that transcendental numbers did, and the fact that π was the solution of a transcendental equation like $\cos(x) = -1$ did not mean that it still might not appear as the solution of an algebraic equation as well. In fact, at the turn of the nineteenth century, mathematicians could not point to a single irrational and say 'this number is transcendental'.

The existence of transcendental numbers was first definitely established in 1844 by the French mathematician Jacques Liouville. He, in turn, went on to actually find one. It was a curious, and in all other respects not overly significant, number starting

$$0.11000100000000000000000001000\ldots$$

with zeroes everywhere except in the first (1), second (1×2), sixth (1×2×3), twenty fourth (1×2×3×4), one hundred and twentieth (1×2×3×4×5) and so on decimal places, where it had a one. From this beginning, Liouville went on to find whole classes of numbers which were not the solutions of any algebraic equations. Unfortunately pi was not one of them. Finally, however, nearly forty years later (in the year 1882 to be precise) the German mathematician Lindemann succeeded in showing that pi was, as long suspected, a transcendental number. After two thousand years, efforts to square the circle could therefore finally be put aside.

If transcendental numbers are so difficult to find, you may be tempted to ask, why do we need to worry very much about them? Well, astonishing as it may seem, it is now known that nearly all numbers are transcendental. By this we mean that if we took our (mathematically ideal) piece of string of unit length, and cut it quite arbitrarily into two parts, then the chances that the two resulting pieces had lengths which could be measured in algebraic numbers are essentially nil. We shall discuss this fascinating point again in Chapter 21 but, for the moment, all we need to know is that the transcendentals are out there in great numbers. It is our system of counting which is just not easily adapted to look for them. Like all the algebraic irrational numbers, they appear as an infinite decimal sequence of non-cycling digits and, to put it bluntly, one set of never-ending patternless digits looks very much like another.

As an interesting example of how ubiquitous the transcendental numbers are let us return to the transcendental equations like $\sin(x) = 1/2$ and $\ln(x) = 2$ set out above. Not only do both these specific examples have transcendental solutions, but the more general forms

$$\sin(x) = y$$

and

$$\ln(x) = z$$

have transcendental solutions x for *all algebraic values of y and z* except for the cases $y = 0$ and $z = 1$ referred to above (for which $x = 0$). Think about this for a moment. It means that if we plot the graphs of $y = \sin(x)$ or $y = \ln(x)$ on a piece of paper, each one goes through only a single algebraic point in the entire infinite x, y-plane (where by an algebraic point we mean one for which both coordinates x and y are algebraic numbers). Since algebraic points are infinitely dense in the plane, these curves both accomplish the remarkably difficult feat of winding between this infinite number of algebraic points without touching any of them. But maybe our amazement can now be tempered by the knowledge that transcendental points in the plane are even infinitely more concentrated than their algebraic counterparts. A point chosen at random in the plane (with a mathematical 'pin') will hit a transcendental point with virtual certainty. It is just a quirk of our counting system which seems to persuade us otherwise.

18. MOST NUMBERS ARE NORMAL, BUT IT'S TOUGH TO FIND ONE

If you look at a long run of digits in a number like pi or the square root of two, it seems as though each numeral occurs about as frequently as any other. Lacking any specific reason for supposing otherwise, one would not therefore be unduly surprised if a careful statistical analysis of, say, the first one million digits of pi showed that there were about 100,000 ones, 100,000 twos and so on. This does, in fact, appear to be the case. Thus, although the numeral 1 (for example) appears in pi in a most unorderly fashion (showing up in the 1st, 3rd, 37th, 40th, 49th, 68th, 94th, 95th, ... decimal places) it does appear approximately a fixed fraction of the time on average, and this fraction is, of course, one tenth. For the other numerals this same overall frequency of appearance also seems to occur, although there *are* some unusual patterns of digits in the first one million decimal places of pi, such as the seven consecutive threes which follow the 710,099th digit.

Now suppose that we are a little more ambitious and look for the frequency with which a particular two-digit pair of numerals appears in pi. Would we also find that an arbitrarily chosen pair, like 79 for example, appears on average once in every one hundred digits? Well, the combination 79 actually occurs twice in the first thirty decimal places, but then it 'settles down' to occur only seven more times in the first one thousand digits, from where it does seem to follow the expected frequency on average. What if we continued the exercise to look at arbitrary 3, 4, 5, ... digit numeral patterns; would they occur on average once every 10^3, 10^4, 10^5, ... digits? Again, lacking any specific reason for believing otherwise, we would perhaps not be surprised if this was the case.

Numbers having the property that *every* sequence of n numerals (with $n = 1,2,3,4,5,....$) occurs on average exactly once in every 10^n digits are defined as being *normal*. It is immediately obvious that no rational number, with its cycling digit pattern, can ever be normal. But what about the irrational numbers with their patternless and boundless sequences of non-recurring decimals? They certainly could be; but how many actually are? Hopefully quite a lot because, unless a goodly number of them are, the choice of the word 'normal' to express this property would hardly be appropriate. Could *all* the irrational numbers perhaps be normal? With a little thought we can certainly rule out this possibility, since it is no trouble to define an infinite patternless decimal which has some numeral (say 9) completely missing. Such a number is irrational and it certainly exists and yet, with the probability of finding a 9 exactly equal to zero, it cannot possibly be normal. But surely, I hear you say, that sort of number is a bit artificial and is not likely to arise very often by chance. What about commonly used irrationals like $\sqrt{2}$, $\sin(1/2)$, $\ln(2)$, and π; are they normal? Sad to say we do not yet know for certain because, although they give every indication of being normal in terms of numeral distributions over their known decimal expansions (upwards of a million digits in some cases), the property of 'normality' can never be proved by hunting digit-patterns with a computer, since we can never 'get to infinity' to check out the probabilities *exactly*.

At the time of writing it is not known whether *any* of the common irrational numbers of elementary mathematics are normal. Could it be then that we have really deluded ourselves? Maybe no number is normal in its exact sense. This thought may also happily be dismissed since, believe it or not, it is extremely simple to invent one. It is a rather preposterous number to be sure, but certainly normal according to our definition. The number begins

0.123456789 10111213141516171819 20212223242526272829 3031......

and is obtained by writing the concecutive integers 1, 2, 3, 4, ..., 10, 11, 12, 13, ... in a row and putting a decimal point in front. This number has been proven to be not only irrational but transcendental as well. Yet, interestingly, although it is normal in decimal notation, no-one yet knows whether it is normal when expressed in any other base. Unfortunately, examples of numbers which are demonstrably normal are extremely difficult to come by. Given this difficulty it is surprising to find out that a general proof exists which establishes that *almost all numbers are normal,* even though we cannot locate many of them with certainty. By this we mean, using our (mathematically ideal) unit-length piece of string once again, that we can find a sequence of intervals of total length as small as we like in which *all* of the non-normal numbers less than 1 are found.

Let us think what this means; and the implications are rather astounding. It means, for example, that if pi is a normal number, then somewhere in its decimal expansion (if we look far enough) we shall find *any* sequence of digits we care to name and, moreover, we shall find it not once but infinitely often. Now it is not difficult to translate any English text into a numerical code by assigning numbers to the letters of the alphabet and the punctuation marks etc. Suppose that we choose one particular code and translate any book (say the Bible or the Complete Works of William Shakespeare) using it. We finish up with an extremely long sequence of precisely determined digits - and this exact same sequence occurs not only in the decimal expansion of every normal number, but it occurs infinitely often. What is more, so do equivalent translations in any other arbitrarily chosen number-code. All the secrets of the universe are here, discovered and undiscovered, and in countless coded forms. Of course, in practice one can never extract any of this information, since the length of decimals needed to find even the simplest of coded statements would be unimaginably large. Nevertheless, in principle at least, all of this is implied by the simple concept of a normal number, and most numbers are normal.

Having defined a normal number it is perhaps interesting to think a little more about the meaning of 'randomness'. Deep inside we all believe that we know what random means, and the concept of a normal number seems to provide examples in which successive numerals appear in a random fashion. Yet how can the successive digits of pi, for example, possibly be random when, via the series relationships of the last chapter, we now possess well-defined methods for extracting them. Surely a random number is one which no gambler, however intelligent and well prepared, can possibly do better than have a one-in-ten chance of guessing the next digit. If our super-intelligent gambler eventually recognized a number pattern as pi, or $\sqrt{2}$, or any number which could be generated in a well-defined fashion, he would get all of the subsequent digits right even if the number were normal. It is clear then that a truly random number implies something going beyond the concept of normal. Are there any numbers which are random in this stronger sense?

Before deciding this question we must first confront the problem of deciding how we recognise randomness. Consider the two sequences of digits below:

$$123123123123123123123123,$$
$$298859200066702864083897.$$

The first is obviously constructed according to a simple rule, and none of us would be tempted to think of the digits as random. The second was obtained (in a little experiment conducted at my writing desk) by taking balls numbered 1,2,3,4,5,6,7,8,9,0 and picking them in turn out of a hat, returning each to the hat before making the next choice. The latter procedure might well be acceptable to most as a method for generating random digits, providing that the hat was given a long enough shake between choices. Yet, in fact, according to the laws of probability, the first sequence of digits above had exactly the same chance of being pulled from the hat as the second (to be precise, one chance in 10^{24}, since there are 24 digits in each series and any could have been one of ten equally probable numerals). From this point of view the first series should qualify as random just as much as the second, and this conclusion is hardly helpful in distinguishing the random from the non-random.

Clearly a better definition of randomness is needed, one which does not contradict the notion of 'lack of orderliness' which we tend to associate with randomness. The task of finding one is by no means easy and only quite recently has such a definition been devised. Essentially it is this; a number is random if it cannot be described any more compactly than actually giving the digital form of the number itself. In this sense the first number above is not random because it can be described as '123 repeated 8 times' while there is no way (at least no way evident to me) that I can describe the second number more compactly than just giving its digits in order. Random numbers are therefore numbers which have nothing special about them; nothing that will enable them to be described concisely. This definition of randomness seems eminently sensible, but unfortunately it does lead to something of a dilemma when we apply it to decimal expansions of irrational numbers. The dilemma is that these expansions go on forever and contain an infinite number of digits. They are therefore only random in the strong sense if no instructions for generating them can be set out in a finite number of steps. This means that we can never calculate a truly random irrational number for, if we can devise a method of calculating it *even in principle,* then it falls outside the definition of random. Yet, astonishingly, it can be proven that nearly all irrational numbers must be random in this strong sense as well. So here we reach the final frustration; nearly all numbers contain a truly random boundless sequence of digits, but it is impossible to devise any method for computing a single one of them!

This finding is singularly unhelpful from a practical point of view. But all is not lost. For a *finite* sequence of integers (which in practice is all that we can ever actually write down) the situation is much more encouraging. A sequence may be judged random if it is 'incompressible', in the sense that a rule for generating it (such as a set of instructions to a computer) cannot be given which is substantially shorter than the series itself. If there is anything at all about the sequence which would give our super-intelligent gambler a better than one chance in ten of guessing each subsequent digit, then randomness has not been achieved. In practice, of course, deciding this question for any particular sequence of digits may not be easy.

Having found that most irrationals are normal with, among other things, one tenth of their digits zeroes, one tenth ones etc., let us now consider the infinite series of counting numbers 1,2,3,4,5,.... in this same context. We can't expect the integers to be normal in quite the same sense as most of the irrationals, because each has only a finite number of digits in it, whereas each irrational had an infinite number. Nevertheless, if we look at some extremely large integers picked at random, we sense that the numerals are once again fairly uniformly distributed on the average. Moreover, there is a tendency to feel that this surely must be the case, at least in some statistical sense, and that it only requires the mathematicians to make the statement precise and to find a suitable proof. We are guessing therefore that 'most' integers, with some suitable definition of 'most', have one tenth of their digits equal to to zero, one tenth equal to 1, and so on all the way up to 9. There is a slight problem in that the first numeral in a conventional integer can never be a zero, but so long as the numbers considered are large enough, this initial-digit effect can be made as small as we please.

In order to pursue this problem seriously we must first digress and examine the idea of an infinite series of numbers a bit more precisely. Consider, for example, the result of summing all the fractions

$$1 + 1/2 + 1/4 + 1/8 + 1/16 + ... ,$$

in which each term is just one half of its predecessor, and where the series continuation (indicated by the dots) goes on forever. What do all these terms add up to? The secret in summing this series (if you don't know an algebraic formula for it) is to notice that at any point in the series the following term always takes the sum half of the remaining distance to 2. A pictorial representation of the series is shown in Figure 6. By taking more and more terms one can approach 2 as closely as desired but without ever quite getting there. It is therefore a fairly natural step to think of the limiting sum, if we could take an infinite number of terms, as being exactly equal to 2. In mathematical terms we say that the series *converges* to 2. The above is therefore an example of a series which contains an infinite number of terms but which converges to a finite sum, namely 2. In order to get a finite sum

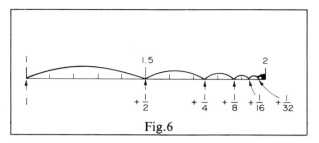

Fig.6

it seems that each term in the series should be smaller than the preceding one, since the sum

$$1 + 1 + 1 + 1 + 1 + \ldots$$

obviously gets forever larger. But the condition that each term be smaller than its predecessor is not sufficient to ensure the convergence to a finite sum. Consider the series

$$1 + 1/2 + 1/3 + 1/4 + 1/5 + 1/6 + 1/7 + 1/8 + \ldots$$

where the dots again imply that the series goes on forever. Does this series have a finite sum? At first glance there appears to be no obvious reason why it should not, but with a little thought we can easily see that it doesn't. For example, let us group the terms as follows:

$$1 + (1/2) + (1/3 + 1/4) + (1/5 + 1/6 + 1/7 + 1/8) + \ldots \,,$$

where the brackets contain respectively 1, 2, 4, 8, 16, 32, ... terms. We now note that the first bracket is equal to $1/2$, while the second one is bigger than $(1/4 + 1/4) = 1/2$, and the third bracket is also bigger than $(1/8 + 1/8 + 1/8 + 1/8) = 1/2$. In fact, each and every bracket is larger than the number of terms which it contains times the last term in it, and this product is *always* exactly equal to $1/2$. In this way we have now established that the series in question is larger than

$$1 + 1/2 + 1/2 + 1/2 + 1/2 + 1/2 + \ldots$$

and this series obviously grows and grows without limit as we continue to add terms ad infinitum. It follows that the series

$$1 + 1/2 + 1/3 + 1/4 + 1/5 + 1/6 + 1/7 + 1/8 + \ldots$$

has a sum which exceeds any number which you care to name. We say that it *diverges,* and we shall write this particular series in the shorthand form Σ $(1/n)$, where the Greek letter Σ symbolizes addition.

By now you may be wondering what all this has to do with the distribution of digits in integers. Well, our problem is that there are altogether an infinite number of integers, and we shall wish below to split them up into different groups according to their digit distribution and to ask which groups are larger or smaller than others. Mathematicians actually prefer the word *sets* to groups, reserving the word group for a rather special collection of numbers with particular properties which do not concern us here. Perhaps, therefore, we had better conform and re-word our problem. Since there are to be an infinite number of integers in each set, how can we possibly decide which set is bigger than another?

In Chapter 21, where we discuss the problem of counting infinities, we shall see that not all infinite quantities are alike. Some infinities are undoubtedly much larger than others, and it is in this sense that we have been able to make statements earlier concerning the fact that there are far more (in fact infinitely more) irrational numbers than there are rational ones, and more random numbers than there are non-random ones. However, for the complete set of natural numbers 1, 2, 3, 4, ... , which are referred to as being *countably infinite,* we are confronted with a different kind of problem, since any two infinite subsets of integers involve the same sort of infinity and are equal. On the other hand, for arbitrarily large, *but finite,* sets of integers (which we might refer to as 'almost infinite') the relative sizes of the subsets can certainly differ, and we must resort to some other method for deciding which is the larger. One possible way is to look at the series Σ $(1/n)$. We have seen above that if *all* the integers are included in n then this sum diverges. On the other hand if we include only some of the integers, as in the sum

$$1 + 1/2 + 1/4 + 1/8 + 1/16 + ... ,$$

then the sum can quite possibly converge. Let us describe an 'almost infinite' set of integers as being 'large' if the series Σ $(1/n)$ tends to diverge when these numbers are substituted for n in the series, and 'small' if it tends to converge. This seems to be quite sensible, and with this definition of large and small some interesting results have been obtained concerning the digit distribution in large integer numbers. For example, the set of integers which has any particular numeral completely missing is small. This is what we should expect. Can we perhaps show that a set of integers must be small if it does not contain about one tenth of each of the numerals 1,2,3,4,5,6,7,8,9,0? If we can, then it would be correct to say that 'most' integers have a uniform distribution of numerals on the average. This problem has been studied by mathematicians with the following rather puzzling conclusions.

Firstly the pleasing part. In accord with our intuition, the set of numbers in which any *specific* numeral occurs in one tenth of the available digit locations is indeed a large set. Moreover, the set of integers with any two specified numerals each occurring with a frequency of one tenth is also large. But now for the puzzling part. If we specify three or more numerals, then it is found that the corresponding set of numbers in which each occurs with a probability of one in ten is *small*. This is equivalent to saying that three or more numerals do not occur together the 'expected' number of times very often. For example, most integers have one tenth of their digits 0; also most integers have both one tenth of their digits 0 and another tenth of their digits 1; but (incredibly) rather few have a tenth each of 0, 1, and 2. The conclusion which we are forced to draw, quite against our intuition, is that most integers do *not* have their numerals evenly distributed at all. Whether this 'disconcerting' finding can be attributed to our peculiar definition of 'large' and 'small' in the context of 'almost infinite' subsets, or whether it does involve an inescapable truth, remains to be seen.

19. A DIFFERENT WAY OF COUNTING; GEOMETRIC NUMBERS

Readers who have progressed this far in the book are by now familiar with the shorthand 'exponent' notation used for many tens multiplied together. We have found it convenient to write one million (1,000,000) as 10^6 and one billion (1,000,000,000) as 10^9. Although this scheme is less of a shorthand when we get down to smaller numbers like $100 = 10^2$ and $10 = 10^1$ it is still nonetheless perfectly consistent and is really only another way of counting. We shall refer to it as counting *geometrically* (in contrast to the common system used in everyday experience which now, in order to distinguish it, we shall refer to as counting *arithmetically*). In geometric counting we are therefore already familiar with the meaning of the positive integers as follows:

Geometric Counting	Equivalent Arithmetic Number
1	10
2	100
3	1,000
4	10,000
5	100,000
6	1,000,000

and so on. The geometric integers count the number of arithmetic tens to be multiplied together.

What, you may be asking, is the use of that? Most of the common counting numbers like 23 or 586 are missing from the arithmetic column altogether and so are not counted at all in the geometric system. But wait a minute; so far we have defined only the geometric *counting numbers*. The arithmetic number system, which we developed in the earlier chapters, also started with positive integers. It was then gradually expanded to include first a zero, then negative integers, then fractions, and so on to the irrational numbers. Who is to say that we cannot do this in a perfectly consistent manner for geometric numbers as well, after which there may be an exact correspondence between *every* geometric number and its arithmetic equivalent? But how can we move forward to the concepts of a geometric zero, negative integer, fraction, and so on? First we notice a useful property of the geometric counting numbers as follows: when we multiply two of the arithmetic integers together from the right hand column in the table above, the answer can be obtained by simply adding together the corresponding geometric numbers. To be specific, the arithmetic result

$$100 \times 10{,}000 = 1{,}000{,}000$$

can be rewritten as

$$10^2 \times 10^4 = 10^6$$

and therefore, in terms of geometric integers, becomes

$$2 + 4 = 6.$$

Suppose that we *define* geometric numbers to be the numbers which satisfy this adding condition whenever *any* two arithmetic numbers are multiplied together. If this is the case, then we can immediately begin to expand the concept of geometric numbers consistently from the positive integers to the more general cases. What, one might ask, is the meaning of the geometric zero? Since

$$100 \times 1 = 100$$

is true, we must have $1 = 10^0$ for then, and only then, will this arithmetic equation translate into a 'power' form

$$10^2 \times 10^0 = 10^2$$

in such a way that the exponents (which are the geometric numbers) obey the required 'additive rule' $2 + 0 = 2$.

Extending this idea a little further, it must also be true that $1/10 = 10^{-1}$, since then the relationship

$$100 \times (1/10) = 10$$

translates to

$$10^2 \times 10^{-1} = 10^1$$

and the geometric numbers again work out to be additive in the form $2 - 1 = 1$. Similarly, as is now easily verified, we must also have $1/100 = 10^{-2}$, $1/1000 = 10^{-3}$, and so on. We have therefore now learned how to count geometrically using both positive and negative integers and zero. More accurately, we have not learned it, but have worked it out for ourselves from a single 'additive rule' which defines geometric numbers in general. Hence we can extend our explicit comparison of geometric and arithmetic numbers further as shown in Table 12.

TABLE 12	
Geometric Counting	Arithmetic Equivalent
-6	0.000001
-5	0.00001
-4	0.0001
-3	0.001
-2	0.01
-1	0.1
0	1
1	10
2	100
3	1,000
4	10,000
5	100,000
6	1,000,000

So far so good; but we still have to move on to consider geometric fractions and irrationals. Can we also do this with our 'addition rule'? Yes we can, and it isn't even difficult. For example, we all know that $1/2 + 1/2 = 1$. If this is read as a geometric number statement it must mean, in regular (arithmetic) numbers

$$10^{1/2} \times 10^{1/2} = 10^1 = 10.$$

It follows that the geometric number $1/2$ is just the arithmetic number which, when multiplied by itself, makes 10. This is the number we are more familiar with as the 'square root of ten'. Pushing ahead still further, since

$$1/3 + 1/3 + 1/3 = 1,$$

it must also be true that

$$10^{1/3} \times 10^{1/3} \times 10^{1/3} = 10^1 = 10$$

so that the geometric number $1/3$ is the regular number which, when multiplied by itself, and then by itself again, makes ten. This we usually call the cube root of ten. By simple extension we can now understand the meaning of all the geometric fractions in the sequence $1/2, 1/3, 1/4, 1/5, \ldots$ and so on. They correspond to the arithmetic numbers which we call the square root, cube root, fourth root, fifth root, etc. of ten. In the 'arithmetic world' these numbers are all irrationals (with explicit forms 3.16227..., 2.15443..., 1.77827..., 1.58489... and so forth), in the geometric world they are rational.

The next extension is to fractions which do not necessarily have a 1 on the top. Consider the geometric number 2/3 for example; can we deduce its meaning as well? We certainly can; and by the same method. Since $2/3 + 2/3 + 2/3 = 2$, it follows from our 'addition rule' that we must have

$$10^{2/3} \times 10^{2/3} \times 10^{2/3} = 10^2 = 100,$$

so that $10^{2/3}$ is just the cube root of one hundred (which is the irrational number 4.64158...). It is now not difficult to see the generalization valid for all geometric fractional numbers. The geometric fraction n/m stands for the arithmetic number $10^{n/m}$, which is the mth root of 10^n or, equivalently, the number which when multiplied by itself m times makes 10^n. The extension to negative fractions is equally apparent. Since

$$-1/2 - 1/2 = -1$$

it must hold that

$$10^{-1/2} \times 10^{-1/2} = 10^{-1} = 1/10.$$

It follows that the geometric number $-1/2$ corresponds to the arithmetic number which, when multiplied by itself, makes 1/10 (in other words to the square root of one tenth). Specifically this number is 0.31622... in decimal form.

With the single 'additive rule' we have now been able to construct a whole new rational number system, the geometric numbers. The extension to irrationals is accomplished exactly as it was for the normal arithmetic numbers. Recognizing that we can never write down an irrational number in a finite number of decimal places, we must be content to represent it by a rational fraction chosen as close to it as we wish to get. What, for example, is the geometric number $\sqrt{2} = 1.41421...$ equal to in regular arithmetic numbers. We can answer this question to any chosen degree of accuracy, but not exactly. Since $\sqrt{2}$ is a little larger than 1.4 (or equivalently 14/10) the arithmetic number it stands for must be a little bit larger than

$$10^{14/10} = 25.1188....$$

It must also be smaller than

$$10^{142/100} = 26.3026...,$$

since the fraction $142/100 = 1.42$ is a bit larger than $\sqrt{2}$. In this way, by choosing fractional exponents closer and closer to $\sqrt{2}$, we can finally zero-in on its arithmetic equivalent which is (to four decimal places) 25.9545...   .

What are the use of these geometric numbers now that we have them? Surely one way of counting is quite enough. Well, the geometric numbers are obviously extremely useful when it comes to multiplying regular numbers, since it is only necessary to add the geometric-number equivalents, and adding is a lot simpler than multiplying. Some of my older readers may now feel that they have some vague recollection of doing this sort of thing in far-off school days. Isn't this something to do with logarithms? Indeed it is; the geometric numbers are in truth just those wretched logarithms which (before the era of the pocket calculator) were often such a frustrating part of our computational education. If I had called them logarithms in the first place, I would no doubt have lost most of my readers at the chapter's outset (since the notion persists that logarithms are somehow difficult except for the expert), so I carefully renamed them 'geometric numbers' to avoid prejudice. Nevertheless the fact remains that the geometric number for $\sqrt{10}$ (or 3.16227...) which we found above to be 1/2 (or 0.5), is just the logarithm of $\sqrt{10}$; that is log(3.16227...). This can readily be verified (if anyone these days has an old set of log tables) by looking up the log of 3.162 and finding 0.500. For the modern generation, no more work is necessary than pressing the appropriate log button on your pocket calculator.

To be rather more specific, the geometric numbers of this chapter are really the *logarithms to the base 10,* since they are based on powers of 10. Geometric numbers (or logarithms if you prefer) can be equally easily based on *any* other arithmetic number, rational or irrational, but base-ten logs were 'normal fare' in most sets of log tables. However, there is one other base which is extremely important for geometric numbers, and we can approach it as follows, although the full power and usefulness of the concept of *natural logarithms,* which it spawns, is wrapped in the mysteries of integral calculus and is beyond the scope of our present story.

Let us think back a moment to our definition of the geometric number zero in base 10. It arose from the equation $10^0 = 1$, which was necessary to satisfy our 'addition rule' for geometric numbers. Geometric numbers a little larger than zero (like 0.0001, for example) must evidently represent arithmetic numbers which are a little larger than 1. When checked out numerically it is found that, for *any* small number x, the geometric number corresponding to $1+x$ is 0.4343x (or more exactly 0.43429448...x). By this we mean that

$$10^{0.4343x} = 1+x$$

for any value of x which is small enough. In logarithmic language this translates to

$$\log(1+x) = 0.4343x$$

and it follows that, in this 'small x' limit, the logarithm of $(1+x)$ is exactly *proportional to* the number x itself. But what is this 0.4343 on the right hand side of the equation? Wouldn't it be a lot easier if $\log(1+x)$ was just equal to x without this 0.4343 complication. It surely would, and this can actually be arranged if we are only willing to give up the base ten for calculating our geometric numbers. Which particular base would allow us to write the simple relationship $\log(1+x) = x$ for all small values of x? Mathematicians call this very special number e and write

$$\log_e(1+x) = x,$$

calling this particular kind of geometric number a *natural* logarithm. Since the natural logarithms are so useful in calculus, and appear so frequently in more advanced mathematics, it is rather cumbersome to have to keep on writing a subscript in the form $\log_e(1+x)$, and so mathematicians usually use a shorthand notation $\ln(1+x)$ in its place for natural logs. We have seen this form earlier in the book and perhaps now have a somewhat better idea of what it really stands for.

But what is this number e, you ask? How big is it? Is it rational? It can be shown that e is an irrational number and it is, next to pi, the most commonly used irrational number in all of mathematics. Let us try to find out just how big it is. The equation $\ln(1+x) = x$, which defines it, states that the geometric number x corresponds to the arithmetic number $1+x$ when we are working in base e rather than 10.

This means that the number e raised to the power x makes $1+x$ (always provided that x is sufficiently small). In symbols this takes the form

$$e^x = 1+x.$$

Let us rewrite the small number x as $1/n$ where n is a very large integer. Our equation now looks like

$$e^{1/n} = 1 + 1/n$$

or, multiplying each side by itself n times,

$$e = (1 + 1/n)^n.$$

With this equation we can now actually go ahead and calculate e by evaluating $(1 + 1/n)^n$ for larger and larger values of n. Let us try it; we can make a table of the results as n gets bigger and see if e is converging rapidly enough for us to evaluate it to several decimal places. Our results are shown in Table 13.

TABLE 13			
n	$(1 + 1/n)^n$	n	$(1 + 1/n)^n$
1	2	500	2.7156
2	2.25	1,000	2.7169
5	2.4883	2,000	2.7176
10	2.5937	5,000	2.7180
20	2.6533	10,000	2.7181
50	2.6916	10^5	2.718254646
100	2.7048	10^6	2.718281828
200	2.7115	10^7	2.718281828

We see that for n equal to one million we have reached an accuracy of at least nine decimal places. At this stage the value of e is 2.718281828 and the decimal representation appears to be cycling. But, alas, it is only an illusion, and in more detail the decimal representation of e continues as

2.7182818284 5904523536 0287471352 6624977572 ...

and has be been shown to be both irrational and transcendental. The latter, you will recall, means that it can never appear as the solution to an algebraic equation. The best known simple rational approximation for e is $2721/1001 = 2.7182817...$ accurate to six decimal places.

Thus, both π and e are irrational and transcendental. One would therefore suppose that quantities like $e + \pi$ and $e \times \pi$ are also irrational and transcendental. Surprisingly, the irrationality of the latter two expressions has never been established, although it is known that at least one of them must be. It is tempting to suppose that any two numbers with an infinite non-repeating decimal expansion should also have a similar irrational form when added together, but this is not necessarily true. For example, the difference between e and 10 is certainly irrational and transcendental. It is a number which starts as 7.2817181715.... and goes on forever without repeating or cycling. Nevertheless it is quite obvious from its definition that when added to e it results not only in a rational sum, but actually gives an integer, ten.

It seems extremely unlikely (but, until someone proves otherwise, it is just possible) that pi and e could be related by some unknown formula which would give their sum as a rational number; that is as a repeating or cycling decimal with a very long period. It would have to be an extremely long repeat distance indeed since both pi and e are already known as decimals to a vast number of decimal places and no sign of cycling has yet appeared when they are added together. Other curious objects like

$$e^e, \quad \pi^\pi, \quad \pi^e,$$

have also not yet been proven to be irrational, although it has been established recently that e raised to the power π is both irrational and transcendental. In all probability they are all transcendental, but it is always dangerous to rest a mathematical case on probabilities and so we must, for the moment at least, just leave the door open a crack and acknowledge that number theory has surprised us many times before.

Before leaving the mysterious number e we ought to give one more thought to its definition. Although I glibly sketched out the numerical findings for a calculation of $(1 + 1/n)^n$ as n got bigger and bigger, I was able to do so only by virtue of having a fairly sophisticated pocket calculator by my side. You can see from Table 13 that it takes powers of the order of one million in order to obtain an answer correct to about ten significant figures. Can you imagine what a task this would have been without my calculator? Try multiplying 1.000001 by itself even a few hundred times for recreation! How then did mathematicians calculate e accurately before the computer and calculator era? The answer lies in an ability to express a quantity like $(1 + 1/n)^n$ as a *power series*. By examining in turn $(1 + 1/n)^2$, $(1 + 1/n)^3$, $(1 + 1/n)^4$, etc. the pattern which the expansion takes soon becomes obvious to anyone who has had a little high school algebra. Mathematicians call the resulting terms the *binomial expansion* and its origin goes way back over the centuries. Explicitly it takes the form

$$(1 + 1/n)^n = 1 + n(1/n) + [n(n-1)/1\times2](1/n)^2$$

$$+ [n(n-1)(n-2)/1\times2\times3](1/n)^3 + \cdots$$

Using this expansion in the limit of n going to infinity, one finds the result

$$e = 1 + 1/1! + 1/2! + 1/3! + 1/4! + \ldots$$

where the exclamation mark behind an integer is the factorial we met in Chapter 4, and means a multiplication of that integer by all smaller ones down to 1 in turn. For example, $1!=1$, $2!=2$, $3!=6$, $4!=24$, and so on, so that the series for e can also be written as

$$e = 1 + 1 + 1/2 + 1/6 + 1/24 + \ldots$$

although we usually prefer the (!) notation since it exhibits the pattern of continuation implied by the dots in a clearer fashion. This infinite series for e, which may well be the most important series expansion in the whole of mathematics, was discovered by Isaac Newton and first appears in a famous treatise written in 1665.

The advantage which this series has over most of its π counterparts is that it is both extremely simple *and* converges very rapidly. Factorial 10 (10!) is already in the millions, and these numbers appear on the bottom of the fractions to be added, so that the latter get rapidly smaller. Using the series expansion for e, we do not need to know anything like a million terms in order to obtain the first ten significant figures of e. How many terms do we need? Let us set them out and see:

number of terms	value of the series
2	2
4	2.6667
6	2.7167
8	2.718253969
10	2.718281527
12	2.718281828

Just twelve additions gets us the same accuracy that about one million multiplications took before. Although I must admit that I again worked with my calculator, it would not take a conscientious and competent mathematician long to get this far with a pencil and paper if alternatives were not available. This, I think, is an excellent example of the manner in which a little mathematical insight can pay enormous dividends in terms of time and effort saved in computational activity. Even so, rather surprisingly, computer programs for calculating e are not yet as efficient as those available for π or $\sqrt{2}$ when it comes to determining hundreds of thousands of decimal places. Thus, while at this writing the series expansions for both π and $\sqrt{2}$ are known to well beyond the one millionth decimal place, the expansion for e has not yet, to my knowledge, progressed to this millionth milestone.

20. TWO DIMENSIONAL NUMBERS

Let us pause once again to rethink exactly what we mean by number. We recognized at the outset of this book that number is a rather abstract idea, but one which is useful in answering questions like 'how many?', 'how long?', or 'how far?' Through the pages we have increased the degree of sophistication of the concept of number until now, with integers, fractions, and irrationals, we can express the position of any point on a 'number line' at least in principle. Unfortunately this 'linear' system of numbers is still not satisfactory. It is not complete. By using it we can always add, subtract, multiply and divide, but we cannot always form powers or roots. This means that there are still arithmetic questions we can ask which do not have answers in terms of the number system we have so far developed. Look back at Table 12 in the last chapter, for example, and you will see that there are no geometric numbers (or powers) which correspond to negative numbers. We might therefore ask 'what is the geometric number (or logarithm) of -1?' At an even more elementary level we might also ask 'what number, when multiplied by itself, makes -2?' In other words if we search for quantities like $\ln(-1)$ and $\sqrt{(-2)}$ or, equivalently, for solutions of equations like

$$e^x = -1,$$

and

$$x^2 + 2 = 0,$$

they are just not available in the number system so far developed. Our challenge is therefore to extend the concept of number still further, in order to make it complete in all the operations of arithmetic. Since the numbers which measure position along a line are clearly not sufficient, it is not unnatural to look next at numbers which measure position in a plane.

Just suppose, for the moment, that I have a map of a treasure island with a cross on it marking an easily identifiable landmark. Starting from this point as 'ground zero' I can ask the question 'where is the treasure?' I am interested in some part in knowing how far away the treasure is, and it is true that this part of the problem can be answered in terms of the numbers which I already have at my command, but to answer the complete question needs something more. It requires at least two 'linear numbers' plus an idea of relative direction. Thus I want to know how far east (let us say) and how far north. Can we combine these two 'normal' numbers in some way to create a more general kind of number which by itself can tell me where the treasure is?

The task of combining two existing numbers to make a new kind of number is not completely new to us, since we earlier defined fractions in terms of two integers (one divided by the other) in order to broaden the meaning of number from integers to rationals. In the new rational number system the original integers were then retained as special examples. In the two-dimensional (or planar) context, we first let the numbers we already have (or *real* numbers as they unfortunately have come to be called) measure the distances to the east (say). The negative 'real' numbers then obviously measure distances to the west. For measurements to the north and south we evidently need some concept which will distinguish 'north-south' numbers from 'east-west' ones. Suppose, for the moment, that we merely label them differently. Let us write a distance of two units north as $2i$, to distinguish it from a distance of two units east, which we simply label as 2. To add and subtract planar numbers of this kind is not difficult, and the combined number $2+2i$ should mean a distance of two units east followed by a distance of two units north. The 'number-point' reached, labeled by the planar-number $2+2i$, is now neither along an east-west axis nor along a north-south one, but is somewhere out there in two-dimensional space. The fact that the point $2i+2$ (that is going north first and then east) takes us to precisely the same point as before means that $2+2i$ and $2i+2$ are exactly the same planar-number, conforming nicely with our experience in the simpler number systems which have gone before, in which (for example) $2+\pi$ and $\pi+2$ are the same 'real' number.

But what is this i symbol? To investigate more closely we now consider the meaning of multiplication for two-dimensional numbers. It is not immediately obvious what we ought to mean by multiplication in two-dimensions, but surely multiplication by 1 should leave things unchanged in order to conform with our basic understanding of unity. We therefore expect $1\times(2+2i)$ to be the same number as $2+2i$ and, even more simply, that $1\times i$ should equal i. If we now require that multiplication of i by 1 should give the same answer as multiplication of 1 by i (again to conform with the behavior of all our simpler numbers) then the real meaning of multiplication by i suddenly leaps into reality. Thus, multiplication of 1 by i makes i and, in picture language (see Figure 7), this implies rotating the number point from a unit distance east (1) to a unit distance north (i). Multiplication by i is therefore a rotation through ninety degrees in an anticlockwise fashion.

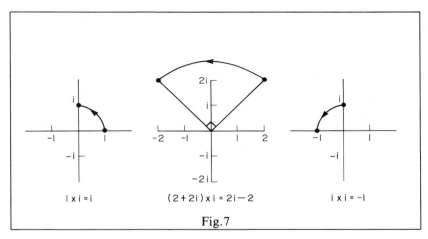

Fig.7

But now suppose that we multiply i by itself. An anticlockwise rotation through ninety degrees from one unit north takes us to one unit west (see Figure 7) and this is the point -1. In other words

$$i^2 = -1,$$

from which we learn that i is the square root of -1. With this result we can now multiply any two-dimensional number by any other, simply by using the rules of algebra. For example

$$(2+2i)\times i = 2i+2i^2 = 2i-2.$$

This operation is also shown graphically in the Figure 7 and verifies once again that multiplication by i is just a rotation anticlockwise through ninety degrees. As another example we might ask what the square of $(1+i)$ is? Using simple algebra, we find the answer

$$(1+i)\times(1+i) = 1\times(1+i) + i\times(1+i)$$

$$= (1+i) + (i+i^2) = 1 + i + i - 1 = 2i.$$

These two-dimensional numbers are usually called *complex numbers* in mathemetics texts, and the number $2+2i$ is said to be made up of a 'real part' equal to 2 and an 'imaginary part' equal to $2i$. This choice of words is particularly unfortunate since there is nothing in the slightest imaginary about i. The confusion comes from the use of the term 'real number' in place of a more proper term such as 'linear number' for quantities like π and $\sqrt{2}$, which measure distances along a line. It is quite true to say that the number i is not contained in our 'linear-number' system; but then the number π is not contained in our rational number system, and the number $1/4$ is not found among the counting numbers; but this is never thought of as 'making them imaginary'. The number i is a two-dimensional or planar number (representing a distance of zero to the east and one to the north) and the planar numbers contain the linear numbers as special cases in just the same way that rational numbers contain the integers.

Although the existence of complex numbers as abstract concepts had been recognized earlier, their geometric representation by points in a plane was first given by a Norwegian surveyor Caspar Wessel in 1797. Unfortunately for him, his achievement went largely unrecognized for a hundred years and a Swiss bookkeeper, Jean Robert Argand, who developed a similar description in 1806, received the credit and still does, in most textbooks, to this day. But it was Gauss, once again, who was the first mathematician to make free and extensive use of complex numbers, and it was he who brought them to full acceptance among the mathematical community at large.

How does all this affect our previous inability to solve equations like $x^2 + 2 = 0$ and $e^x = -1$? Considering the first, we get a clue when we substitute the two-dimensional numbers $\sqrt{2}i$ or $-\sqrt{2}i$ for x in this equation; it is now satisfied. This means that the solutions of the equation $x^2 + 2 = 0$ are planar numbers. More generally, it has long been known that if the highest power of x in an algebraic equation is x^n, then this equation can have no more than n solutions in terms of linear (or real) numbers. For example, the equation

$$x^3 - 6x^2 + 11x - 6 = 0$$

is satisfied by $x = 1$, $x = 2$, and $x = 3$. Check it out for yourself. On the other hand the equation

$$x^3 - x^2 + 4x - 4 = 0$$

has only one solution, $x = 1$, in real numbers. In planar numbers, however, two more solutions can be found to this last equation, namely $x = 2i$ and $x = -2i$. In fact, if the full complex number system is used, it seems that *all* cubic equations have three solutions. More generally still, using these same complex numbers, any algebraic equation in which the highest power term is x^n (what we call an nth-order algebraic equation) is now known to have *exactly* n solutions. This famous theorem, the fundamental theorem of algebra, was first stated by J. le Rond d'Alembert in 1746, but only partially proved. The full proof was given by Gauss (then twenty one years old) in his doctoral dissertation in 1799. Another way of stating this theorem is that any nth-order algebraic equation can be written in the form

$$(x-a)(x-b)(x-c)(x-d)\dots = 0$$

in which a, b, c, d, ... are complex numbers, and where the dots imply that there are exactly n brackets multiplied together in total. The complex number system is therefore just that system of numbers which is needed for the *complete solution of all algebraic equations.*

But what about $e^x = -1$? This is not an algebraic equation; can we solve this equation with complex numbers as well? First let us get a little better acquainted with complex numbers in general. Since $i^2 = -1$, it follows immediately that any power of i can be reduced to $+1$, -1, $+i$, or $-i$. For example, $i^3 = i^2 \times i = -i$; and $i^4 = i^2 \times i^2 = +1$. Any number of complex numbers multiplied together must therefore always result in another complex number. The same is true for division by complex numbers. Division is most easily carried out by noting, using simple algebra, that any complex number $m+in$, when multiplied by $m-in$ (which is called its *conjugate),* results in a real number m^2+n^2. Division by $m+in$ can therefore always be accomplished by multiplying the top and bottom of a fraction by $m-in$ as follows:

$$\frac{1}{m+in} = \frac{m-in}{m^2+n^2}.$$

The result is again just another complex number, represented in our plane by a distance $m/(m^2+n^2)$ east and a distance $n/(m^2+n^2)$ south. It follows that complex (or two dimensional) numbers are completely self-contained as far as addition, subtraction, multiplication and division are concerned. That is, no new kinds of numbers are generated by these operations.

But what about powers and roots? How do we understand what raising a number to a complex power means? How big, for example, is 2^i, or even i^i? Are these complex numbers too? In the language of the last section 'can geometric numbers be complex?' The answer is that they can, and the connection is made via a well-known mathematical expansion for e^x. Since, from the last chapter, the number e is the limit of $(1+1/n)^n$ as n becomes larger and larger, then e^x must be the limit of $(1+1/n)^{nx}$ also for large n. This sort of calculation for real numbers can be performed by using the binomial expansion set out in the previous chapter, and gives the result

$$e^x = 1 + x + x^2/2! + x^3/3! + x^4/4! + \cdots .$$

Complex powers (that is exponents) are defined by taking this equation to be true for all numbers, complex as well as real. Thus, putting x equal to i in the above series, we find

$$e^i = 1 + i - 1/2! - i/3! + 1/4! + \cdots$$

which, summing up all the terms to infinity, converges to the complex number

$$e^i = 0.54030 + 0.84147i$$

to five decimal places. Since, again from the last chapter, $\ln(n)$ is just the base-e geometric number for n, we have $n = e^{\ln(n)}$ and, consequently, $n^i = e^{i\ln(n)}$. In particular, since $\ln(2) = 0.69315...$, it follows that 2^i is the same number as e raised to the power i times $\ln(2)$, or $e^{0.69315i}$. From our series expansion above we now find out that

$$2^i = 0.76923 + 0.63896i ,$$

again a complex number. More generally the value of *any* complex number raised to the power of *any* other complex number can be shown to be always a member of the two-dimensional number set. But there are some surprises. For example, i raised to the power i turns out to be a real irrational number beginning 0.2078795... , while the ith root of i (by which we mean $i^{1/i}$) is also real and irrational with a value 4.8104773.... . These calculations were first performed by Euler, who also noticed that i is very intimately related to the two best-known real irrationals π and e.

Relationships connecting the numbers i, π, and e are based on a finding that e^{ix}, via its expansion

$$e^{ix} = 1 + ix + (ix)^2/2! + (ix)^3/3! + \cdots$$

is directly related to the trigonometric functions $\cos(x)$ and $\sin(x)$. The relationship is a beautifully simple one

$$e^{ix} = \cos(x) + i \, \sin(x)$$

and makes complex numbers of enormous value in trigonometry. Most fascinating of all, to me, is the result which follows when x is put equal to π in this equation. Those of you who are familiar with a little trigonometry will recall that the value of $\cos(\pi)$ is -1 while the value of $\sin(\pi)$ is zero. Substituting these into the above relationship reduces it to

$$e^{i\pi} = -1,$$

an equation which relates those mysterious quantities i, π, and e in a startlingly simple manner. It also provides us at last with the solution to that equation $e^x = -1$; namely $x = i\pi$. In other words $\ln(-1)$ is the complex number $i\pi$.

If the equation $e^{i\pi} = -1$ still seems a bit abstract to you and not too clear, then I can sympathize. It has appeared the same way to many others. But it really is true, and we can bring it to life simply by expanding it in its series form as follows:

$$e^{i\pi} = 1 + i\pi + (i\pi)^2/2! + (i\pi)^3/3! + \cdots$$

$$= 1 + i\pi - \pi^2/2 - i\pi^3/6 + \pi^4/24 + \cdots$$

and plotting it term by term in the north-south-east-west plane. This is done for you in Figure 8 where we observe a counter-clockwise spiral of straight lines which quickly 'strangle' the -1 point on the real (that is east-west) axis. Seeing is believing!

The complex numbers $m+in$ are therefore a complete set with respect to all the operations of arithmetic. Among the subsets of this system we can logically define *complex integers* (when m and n are both integers), *complex rational numbers* (when m and n are both rationals) and so on. In the complex integer system we can now go on to see that some of our 'real-number' primes are prime no longer. For example, we easily verify that

$$(1+i)\times(1-i) = 2$$

and

$$(2+i)\times(2-i) = 5.$$

Does this mean that the concept of prime number completely loses its meaning for complex numbers? By no means. In real numbers we define a prime to be an integer exactly divisible only by itself and by 1.

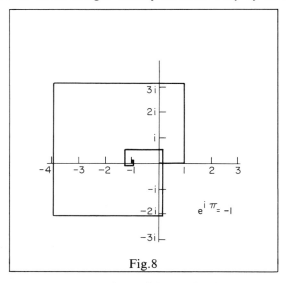

Fig.8

Complex prime numbers are defined by a simple extension which says that a complex number is prime if it is exactly divisible only by itself and by 1 and i. It is necessary to include i in this definition since *all* complex integers are exactly divisible by both i and 1. This is easily proved by noting that

$$\frac{m+in}{i} = \frac{im+i^2n}{i^2} = \frac{im-n}{-1} = n-im.$$

What then are some examples of complex prime numbers? The smallest ones are given in the Table 14 below. Since it can be shown that when $m+in$ is a complex prime number, then so also is $m-in$ and both their negatives, it is only necessary to show those complex primes for which both m and n are positive. A few things can be noted about them. Firstly, if $m+in$ is prime, then so also is $n+im$. Secondly, with the exception of the smallest prime $1+i$, the values of m and n are never both even or both odd. This is because it is easy to establish that all such numbers are divisible by $1+i$. Finally, in all those primes for

TABLE 14				
THE SMALLEST COMPLEX PRIME NUMBERS				
$1 + i$	$1 + 4i$	$5 + 6i$	$5 + 8i$	$3 + 10i$
$1 + 2i$	$4 + i$	$6 + 5i$	$8 + 5i$	$10 + 3i$
$2 + i$	$2 + 5i$	$0 + 7i$	$7 + 8i$	$7 + 10i$
$3 + 0i$	$5 + 2i$	$7 + 0i$	$8 + 7i$	$10 + 7i$
$0 + 3i$	$5 + 4i$	$2 + 7i$	$4 + 9i$	$9 + 10i$
$3 + 2i$	$4 + 5i$	$7 + 2i$	$9 + 4i$	$10 + 9i$
$2 + 3i$	$1 + 6i$	$3 + 8i$	$1 + 10i$	$4 + 11i$
	$6 + i$	$8 + 3i$	$10 + i$	$11 + 4i$

which both n and m are non-zero, the value of n^2+m^2 is a real prime number of the form 'four times an integer plus one'. The latter real primes are therefore no longer prime as complex numbers, since they are made up from the product of $n+im$ and $n-im$.

Picturing complex numbers as two-dimensional numbers in a north-south-east-west plane removes most of their mystery. In this context it is possible to rationalize the disturbing notion that i is the square root of -1. In fact, so direct and logical are the meanings and operations of two-dimensional numbers that it is only natural to ask whether there are three-dimensional numbers as well. Unfortunately the answer is no. In four dimensions, however, if we are willing to slightly modify one of the rules of conventional arithmetic, a rather straightforward generalization of complex numbers is possible. Modern mathematicians are not at all averse to tampering with a rule of arithmetic if the reward is a completely new type of number, so that the extension has been made with enthusiasm.

It was the nineteenth century Irish mathematician William R. Hamilton who first made the breakthrough to 'four-dimensional' numbers. These 'hyper-complex' numbers are called *quaternions,* and are four-part numbers which combine a 'real' number with three 'imaginaries'. However, these new numbers obey arithmetic laws where the order in which two of them are multiplied is important. In real numbers the value of $\pi \times e$ is exactly the same as $e \times \pi$. Even in complex numbers the order of multiplication is irrelevant. This particular property of the more common number systems is called the *commutative law* for multiplication. Only if we give up this law can the extension from two to four-dimensional numbers be made. But all the other laws and operations of arithmetic can readily be defined in the new system and, in fact, the discovery of the sort of algebra obeyed by quaternions marked the beginning of so-called 'modern algebra', in which all kinds of new number concepts and rules of operation can be introduced.

The idea of a fourth dimension is alien to most non-scientists so that we shall not dwell at length on quaternions. Nevertheless, it is not necessary to go to four dimensions to see the problems which arise concerning the order of multiplication; three dimensions is quite sufficient. In two dimensions we interpret multiplication by the number i as an anticlockwise rotation through ninety degrees, and $-i$ as a similar clockwise rotation. The fact that complex numbers obey the commutative law for multiplication is just a mathematical statement of the fact that finite rotations in two dimensions are commutative. For example, place this book down on a table and think of two rotations; say ninety degrees clockwise (a rotation which we shall label R_1) and ninety degrees anticlockwise (R_2). A little experimentation quickly reveals that it doesn't matter whether we do R_1 first and then R_2 (which we symbolize as $R_1 \times R_2$) or whether we reverse the order ($R_2 \times R_1$), the result is the same; we get back to where we started. In the language of complex numbers this is just a demonstration of the fact that $(-i) \times i = i \times (-i) = 1$. By choosing other planar rotations it is not difficult to verify that the order of operation in a plane is immaterial for *any* pair of rotations of the book, even those which do not bring us back to the starting point.

In three dimensions things are quite different. Hold the book horizontally in front of you and let R_1 now stand for a rotation of ninety degrees away from you and R_2 stand for a rotation of ninety degrees to the left. Try performing the two three-dimensional rotations first in the order $R_1 \times R_2$ and then in the order $R_2 \times R_1$. The results, you will see, are quite different. The order of operation is obviously an important property for rotations in three dimensions. The same is true in all dimensions higher than two and we say, in mathematical jargon, that rotations in more than two dimensions are not commutative. It follows that *any* number system which can be constructed in the higher dimensions (such as the quaternions) must take account of this fact in the rules of the game. Therefore, they must all obey somewhat different arithmetic laws from our more common numbers.

21. COUNTING THE INFINITE

In some of the earlier chapters we encountered numbers so large that it was just not feasible to write them down in decimal form, because they would contain more digits than would fill the entire book. Nevertheless, no matter how large a finite number may be, it is always possible to to imagine one still larger, and this process can be continued indefinitely. Colloquially we say "and so on to infinity". But is it possible to get any precise grasp of this 'infinity'? We know that an infinite number of items is certainly larger than *any* finite number, but is there any more which can be said about it? Does it make any sense, for example, to ask whether there are more fractions than integers, or more irrational numbers than rational ones? Clearly there are an infinite number of each, so that we are really asking whether it is possible to find a method for comparing two different infinities to see which is the larger.

The basic problem facing us when we pose these sorts of questions is that we are comparing numbers which we cannot write down in any normal fashion. One reaction might be that we could perhaps make some progress by performing comparisons as follows. We know, for example, that the set of all fractions actually contains the set of all integers. Surely, therefore, it must be the larger of the two sets. This sounds reasonable except for one thing; we are using concepts which are valid for finite sets of numbers and assuming that they still hold for infinite ones. Is the whole necessarily larger than one of its parts when we are talking about infinite quantities? What, for instance, is one half of infinity, or one quarter of infinity? Is it still not infinity? But if it is, you may say, then surely infinity is infinity and there is not much more to be said about it. Fortunately, for the sake of this final chapter, such is not the case and there do exist different sorts of infinite number. In particular, we shall establish below that there are just as many integers as fractions, but that there are infinitely more irrational numbers than either fractions or integers.

In order to accomplish this, it is necessary to go right back to our first statements in Chapter 1 concerning the origins of counting. In a primitive society, with specific counting words for only the first few integers, how would it be possible to decide, for example, whether there were enough spears for all the hunters? The actual numbers involved, say twenty, fifty, or even one hundred, would be quite as meaningless to them as infinity appears to us. Nevertheless, if the only question of importance was one of *relative* size, such as whether the number of spears is larger than, equal to, or smaller than the number of hunters, then the solution is not difficult. All that is needed is to give each hunter a spear, and to see whether there are spears left over when all the hunters are armed, or whether the spears run out before each hunter gets one. In mathematical terms this process is referred to as 'one-to-one correspondence', and the beauty of it is that it can be continued all the way to infinity if a pattern can be set up between matched sets.

It was the German mathematician Georg Cantor who, in the 1870's, first pointed out the relevance of one-to-one correspondence in the search for a measure of infinity. If we can pair the objects of two infinite sets, so that no object in either set is left out as the procedure is continued to infinity, then the two infinite sets are equal. To see how this works in practice, we might take first the two infinite sets containing respectively all the positive integers and all the positive even integers. This is an excellent starting point because the one-to-one correspondence is particularly simple to set up, and because it dispels our earlier feeling that a half of infinity might be smaller than 'all of infinity'. We assemble the two sets of numbers so that each member of one series is just one half of the equivalent member of the other series, starting out as follows:

$$1 \quad 2 \quad 3 \quad 4 \quad 5 \quad 6 \quad 7 \quad 8 \quad 9 \quad 10 \quad 11 \quad ...$$
$$2 \quad 4 \quad 6 \quad 8 \quad 10 \quad 12 \quad 14 \quad 16 \quad 18 \quad 20 \quad 22 \quad ...$$

with the dots implying a continuation to infinity. Each term in one series is uniquely matched with one term in the other and, no matter how far we continue there must, by virtue of the one-to-one correspondence, be exactly the same number of each. It follows from this rule of matching that there are just as many even numbers as total integers. It is usually difficult for people meeting this statement for the first time to accept it, since we are all so used to living in the world of the finite. There is a tendency to say that when all the even integers are 'used up' there will still be half the series of all integers 'left to go'. This is true for any finite stopping point, but we are now talking about going on *forever;* neither the even integers nor the set of all integers ever does get completely 'used up'.

In dealing with the infinite we must therefore be prepared to meet ideas which seem strange to our finite-oriented experience. There is also another point which we must now confront. Since we have already implied that not all infinities are equal, we shall have to become more precise about which infinity we are talking. We need a new symbolism to count infinities and the simplest available seems to be

$$\infty_0, \quad \infty_1, \quad \infty_2, \quad \infty_3, \quad ....$$

where we make use of the well-known symbol ∞ for infinity in a simple subscripted manner. Now Cantor actually used the first letter of the Hebrew alphabet (aleph) instead of the more mundane ∞, and most formal textbooks on these so-called *transfinite numbers* also use Cantor's notation, but ∞ is such a familiar symbol to even most non-mathematicians that I am tempted to adopt it here. The transfinite number which counts the integers is labeled ∞_0 and, as the progressive series of infinities above suggests, it is indeed the smallest such number. All sets of objects which have this number of members are said to be *countably* infinite.

What are other examples of 'countably infinite' sets? Well, we can immediately locate many among the integers themselves, since all the infinite sequences such as the squares, cubes, triangular numbers, etc. can easily be placed in a one-to-one correspondence with the counting numbers. All that we need in order to establish this one-to-one correspondence is a simple rule for relating the numbers to be paired. Thus, for squares, cubes, and triangular numbers, we may start off like

Integers	Squares	Cubes	Triangular Numbers
1	1	1	1
2	4	8	3
3	9	27	6
4	16	64	10
5	25	125	15
6	36	216	21

where the general term (for integer n) is

Integers	Squares	Cubes	Triangular Numbers
n	n^2	n^3	$n(n+1)/2$

and establishes the precise form of the correspondence all the way to infinity. Other infinite sets of numbers, like those exactly divisible by 10, or by 100, or even by one hundred billion, are all countably infinite, and the obvious question becomes 'what infinite sets of objects are *not* countably infinite?'

It is easy to establish that ∞_0 times any finite number is still equal to ∞_0. But suppose that we multiply ∞_0 by itself; surely this would produce a larger sort of infinity, wouldn't it? Surprisingly not, as can be confirmed by counting the number of points in an infinite square array which has ∞_0 points along each edge. Starting from the top left-hand corner we begin this counting process by labeling the points as follows:

$$
\begin{array}{ccccccc}
1/1 & 1/2 & 1/3 & 1/4 & 1/5 & - & - \\
2/1 & 2/2 & 2/3 & 2/4 & 2/5 & - & - \\
3/1 & 3/2 & 3/3 & 3/4 & 3/5 & - & - \\
4/1 & 4/2 & 4/3 & 4/4 & 4/5 & - & - \\
5/1 & 5/2 & 5/3 & 5/4 & 5/5 & - & - \\
- & - & - & - & - & - & - \\
- & - & - & - & - & - & - \\
\end{array}
$$

The general point m/n is labeled by its row number followed by its column number where, for the moment, the slash (/) just serves to separate the two numbers. We now set up the one-to-one correspondence with the natural numbers as follows:

$$1/1 \ \ 1/2 \ \ 2/1 \ \ 1/3 \ \ 2/2 \ \ 3/1 \ \ 1/4 \ \ 2/3 \ \ 3/2 \ \ 4/1$$

$$\ \ 1 \quad 2 \quad 3 \quad 4 \quad 5 \quad 6 \quad 7 \quad 8 \quad 9 \quad 10$$

continuing in a manner which is made obvious by the pattern. The complete infinite square is filled when every point m/n is labeled, where both m and n can take any integer value between 1 and ∞_0. Although I shall not give it here, an expression for the counting number corresponding to the general point m/n can be written, and establishes a precise algebraic form for the one-to-one correspondence. It follows that every one of the points m/n in the infinite square array can be counted by using the sequence of natural numbers. Our surprising conclusion is therefore that this number of points, which we can formally write as ∞_0^2, is exactly equal to the countably infinite set ∞_0. Using a similar argument, it is possible to consider an infinite cubic array to establish that ∞_0^3 is also equal to ∞_0. In fact the proof works for any finite power of ∞_0, so that we can generalize these findings to read

$$\infty_0^n = \infty_0,$$

if n is any non-infinite integer.

In the continuing search for an infinite number larger than ∞_0 we might now ask 'what about ∞_0 raised to the power ∞_0?' This is a perfectly legitimate question but takes us ahead too quickly. Let us first concern ourselves with the simpler question 'what is 2 raised to the power ∞_0?' In finite numbers we know that $2^2=4$, $2^3=8$, $2^4=16$ and so on, and 2^n is certainly a number much larger than n^2 when n becomes large. It therefore seems quite possible that 2^{∞_0} could be greater than ∞_0^2; but how can we go about checking it? Consider two different symbols, say p and q. The quantity 2^2 is then the number of different ways we can write down two symbols either of which can be a p or a q; specifically these combinations are

$$pp, \ qp, \ pq, \ qq.$$

In a similar fashion, 2^3 is the number of ways we can write down 3 symbols, each of which is a p or a q; in detail they are

$$ppp, \ qpp, \ pqp, \ qqp, \ ppq, \ qpq, \ pqq, \ qqq.$$

Generalizing this, the quantity 2^{∞_0} is the number of different ways we can write down ∞_0 symbols, each of which can be either a p or a q. Suppose this infinite series of infinite strings of p's and q's begins

1. ppqpqqpqp....
2. pqqpppqpp....
3. qqqpqqppq....
4. qppqqqppq....
5. pqpppqqpq....
6. qqpqpqppp....
7. ....
8. ....

where we have picked out the sequences in a random fashion and started to match them against the counting numbers. If we continue the matching process all the way up to ∞_0, shall we have included all the possible combinations of p's and q's which represent 2^{∞_0}? At first sight this might appear to be a tough question to answer but, surprisingly, the answer is delightfully simple to establish. Consider that particular combination of p's and q's for which the first symbol is *not* the first symbol in the combination labeled 1, the second *not* the second symbol in the combination labeled 2, and so on down the list to the general nth term, which has its nth symbol *not* equal to the nth symbol in the combination labeled n. From our listing above, this particular combination must begin

$$qpppqp....$$

By its very definition it is not contained in the infinite list set out above no matter how far down we look. For example, if someone suggests that this combination is the 456th on the list, we can immediately respond that it certainly is not since, by its definition, the above combination has its 456th symbol different from that shown in the position 456 on the list.

This extremely simple observation, first pointed out by Cantor, establishes that 2^{∞_0} is a larger infinity than ∞_0. We label it as ∞_1, the second of the transfinite numbers, although from the above proof we have no assurance that there cannot be another infinity (that is transfinite number) which is bigger than ∞_0 but smaller than ∞_1. This is an interesting point to which we shall return below, but for the moment we simply *define* ∞_1 to be equal to 2^{∞_0}. But now for the big step. It can also be established that 2 raised to the power ∞_1 is a larger infinity than ∞_1 (we call it ∞_2), and that 2 raised to the power ∞_2 is a larger infinity than ∞_2, and so on. The general proof is not difficult and it is set out in Appendix 5, for those who are interested. In this way a whole sequence of infinities

$$\infty_0, \infty_1, \infty_2, \infty_3, ..., \infty_n, ...$$

can be built up with the assurance that there is no such thing as a largest infinity of all. There is nothing special about the number 2 which has been used in the generating relationships for the larger infinities; we could equally well have used 3, or 4, or any finite number, and set down similar proofs. Two was used only because the proofs are easiest of all for this case. But our original question asked for the value of $\infty_0^{\infty_0}$. Although the proof for this one is a bit trickier (and we shall not give it here) it is in fact also equal to ∞_1.

We have now acquired enough knowledge to set out some rules for doing arithmetic with the transfinite numbers ∞_n. Adding or multiplying by any finite number does not affect them. Even raising them to a finite power leaves them unchanged. We can write this formally (where m and n are finite numbers) as

$$\infty_n + m = \infty_n$$

$$\infty_n \times m = \infty_n$$

and

$$\infty_n^m = \infty_n.$$

On the other hand if we raise any finite number m to an infinite power, we get the next larger infinity. Note in passing that this implies that Fermat's last theorem (Chapter 11) is easily satisfied if transfinite numbers are allowed for solutions.

$$m^{\infty_n} = \infty_{n+1}.$$

When we are dealing with adding or multiplying infinities together, the rules are that two infinities added or multiplied together give the larger of the two; thus, for example

$$\infty_n + \infty_{n+1} = \infty_{n+1}$$

and

$$\infty_n \times \infty_{n+1} = \infty_{n+1}.$$

If the two infinities involved are the same size, then no change results:

$$\infty_n + \infty_n = \infty_n$$

and

$$\infty_n \times \infty_n = \infty_n.$$

Finally, the rule for raising infinities to infinite powers is that one infinity raised to a smaller infinite power stays the same, while an infinity raised to the same or higher infinite power is increased to a higher level. For example,

$$\infty_n^{\infty_{n-1}} = \infty_n$$

$$\infty_n^{\infty_{n+1}} = \infty_{n+2}$$

and

$$\infty_n^{\infty_n} = \infty_{n+1}.$$

Having located all these transfinite numbers it would now be nice to find actual physical sets of objects which are equal to some of them. The smallest infinity ∞_0 is no problem. We have already seen that it is equal to the number of integers. It is not difficult to establish that it is also equal to the number of rational fractions, since they can all be written in the very same form

$$(1/1), \ (1/2, \ 2/1), \ (1/3, \ 2/2, \ 3/1), \ (1/4, \ 2/3, \ 3/2, \ 4/1), \ ...$$

already discussed in counting the number of points in an infinite square lattice. The most familiar infinite set which has ∞_1 members is the set of all real numbers, rational plus irrational. The proof of this runs exactly parallel to that given for establishing that $2^{\infty_0} = \infty_1$. Thus, if we list all the real numbers (say between 0 and 1) as infinite decimals, we might start

$$
\begin{array}{ll}
1. & 0.38562849.... \\
2. & 0.90340057.... \\
3. & 0.04772834.... \\
4. & 0.02347661.... \\
5. & 0.93371060.... \\
6. & .... \\
7. & .... \\
\end{array}
$$

and try counting them all by placing them in one-to-one correspondence with the natural numbers 1, 2, 3, 4, ... Consider now a real number in which the first decimal place is *not* that of the number labeled 1 and, generally, in which the nth decimal place is *not* that of the number labeled n. This decimal is not contained in the infinite sequence set out above, even if we look all the way down the list to infinity (that is to ∞_0). The total number of decimals is therefore larger than ∞_0. They can, on the other hand, be placed in a one-to-one correspondence with the sequence of p's and q's representing the number 2^{∞_0}, and are therefore counted by ∞_1. Moreover, since the real numbers can also be placed in one-to-one correspondence with the points on a line, it follows that there are also ∞_1 points on a line of any length.

Since, from our set of rules, we know that $\infty_1^2 = \infty_1$, the suggestion is that the number of points in a plane should also be ∞_1. This can be confirmed as follows. Suppose the position of a certain point on the line defining one edge of a unit square is, for argument sake, 0.7346982340.... . We can extract from this number two different numbers by picking out the even and odd decimal places as follows:

$$0.74924.... \text{ and } 0.36830.... \text{ .}$$

Let these numbers, in turn, measure the 'horizontal' and 'vertical' distances of the location of a point inside the unit square as measured from one of its corners. It is clear that this procedure establishes a one-to-one correspondence between a point inside the square and a point along one edge. What is more, every point in the line will define a corresponding point in the square such that none is left over. It follows that the number of points in the square is exactly equal to the number of points on the line; that is ∞_1. In a corresponding fashion, one can also show that there are ∞_1 points in a cube; it is merely necessary to split up the original decimal into three parts, using every third decimal place. Needless to say, it is immaterial which units we use to measure length, so that these results are quite independent of the size of the square or cube. More generally still, it can be shown that there are exactly ∞_1 points along any line, and in any area or volume of any finite size. This fact was first established by Cantor in the year 1877. It came as quite a surprise to him, since he had been convinced (and had for years been trying to prove) that the infinities ∞_1, ∞_2, and ∞_3 distinguished the different orders of space (that is 1, 2, and 3 dimensions). Quite evidently they do not.

From ∞_1 we now pass on to ∞_2. What infinite sets of familiar objects have the size ∞_2? Although we shall not prove it here, it has been shown that the variety of all possible curves which can be drawn on a plane, or in a portion of space, fall into this category. Since all scientific laws are functions of this kind, it is also the number of all possible laws of science. Beyond ∞_2, however, it is difficult to find common collections providing us with examples of the larger transfinite numbers. Perhaps the only way to attempt this is to say something about the concept of subsets. Suppose that I have 3 objects, which I symbolize by the letters A, B, C. The number of different ways in which I can pick zero or more objects from this set is called the number of subsets of the collection. In this case the number of subsets is $2^3 = 8$, namely

A, B, C, AB, BC, CA, ABC, 0.

More generally, if we have n objects, then the number of subsets is 2^n and, in the infinite limit, it follows that a set of ∞_n objects has $2^{\infty_n} = \infty_{n+1}$ subsets. Thus, if we can imagine a collection made up of all the different ways that we can group together one or more of all the possible curves that can be drawn, then this collection has ∞_3 members. It is now clear that we have reached a position somewhat opposite to that experienced by our stone-age hunter in Chapter 1. He had ample objects to count, but few numbers to count with. With our infinite number of transfinites, we now have far more numbers than we can conveniently use, due to our inability to imagine collections of objects large enough to match to the numbers.

One point remains to be discussed. It is the possibility mentioned earlier that there might be transfinite numbers in between those so far defined. Since, as the years passed, no more were ever found, it became customary to assume that no others existed. This was referred to as the *continuum hypothesis*. No further progress was made on this point until 1931, when Kurt Goedel established that there are some statements in mathematics which can never be proven or disproven by logical deductions from the basic rules (or axioms) defining the subject. We discussed this earlier in Chapter 14, and it happens that the continuum hypothesis is one of these unprovable statements. All that can be shown is that this hypothesis is consistent with the rules of mathematics. It can therefore be inserted as an additional 'rule' of mathematics without leading to inconsistencies. On the other hand, as first shown in 1963 by Paul Cohen of Stanford University, it is also possible to construct other self-consistent models of mathematics which do not contain this continuum hypothesis.

As a final exercise we shall ask how many algebraic and transcendental numbers there are. We already know that there are more real numbers than rational ones. However, we have not yet asked whether this is due to a prevalence of algebraic irrationals (like $\sqrt{2}$), which can be obtained as the solutions of algebraic equations, or whether it is due to the transcendentals (like π and e) which cannot. Because of the ease of finding solutions to simple algebraic equations, and the relative difficulty of locating specific transcendental numbers, it is natural to feel that it is probably the former which are the more abundant. Surprisingly, just the opposite is true. It has been shown that the total number of algebraic numbers can be placed in one-to-one correspondence with the integers, and therefore are measured by ∞_0. It is the 'hard to find' transcendentals which number ∞_1 and completely dominate the rest. Considering the fact that it was not until

1844 that such numbers were even known to exist, this finding is rather astounding. The transcendentals are literally everywhere in the 'garden' of real numbers, even if they are extremely adept at hiding themselves.

Given this fact it is surely required of us to find a few more of these elusive creatures. We have already mentioned that π and e are examples and, once this has been established, it is but a small step to show that any non-zero algebraic power of e, such as e^2, e^3, $e^{\sqrt{2}}$ etc., is also transcendental, and similarly for algebraic powers of π. In addition, any algebraic number (except 0 and 1) raised to an irrational or a complex power is also transcendental. Examples are $2^{\sqrt{2}}$, 2^{π}, 2^{e} and 2^{i}. Moreover, since the number e can be related to the trigonometric functions through the intermediary of complex numbers (as we set out in the last chapter), one can also show that the sine, cosine, or tangent of any non-zero algebraic number is transcendental. If we add to this list the logarithms of the algebraic numbers and the angles (measured in radians) whose sines, cosines, and tangents are equal to algebraic numbers, we begin to see that the list of known transcendentals is now really not so small as our earlier concentration on π and e might have led us to believe. On the other hand, since each of the above sets enables us to locate only ∞_0 new transcendentals, it is clear that we still have seen only the tip of the iceberg of transcendental numbers in general.

22. UPDATE (SEPTEMBER 1985)

Mathematically the greatest event to take place since the main text of the book was written is the proof of the so-called Mordell conjecture. Although its importance goes far beyond our immediate concerns (it has in fact been referred to by one of the more excitable professional mathematicians as the 'proof of the century') we are interested only in its implications for Fermat's last theorem of Chapter 11.

In 1922, Lewis Mordell, a British mathematician, conjectured that a certain large class of equations (which we obviously cannot set out in full here) has only a finite number of solutions in terms of rational numbers. One member of this class of equations was of the form $u^n + v^n = 1$, with $n = 3, 4, 5, \ldots$ and so on. A little thought therefore reveals that a proof of the Mordell conjecture would also imply that the Fermat equation $a^n + b^n = c^n$ (see page 96) can have at most a finite number of solutions in whole numbers if n is greater than 2. Although this falls short of a proof of Fermat's last theorem, which states that this equation has no whole number solutions at all, it does take us one important step closer to this widely believed result. It rules out, for example, my statement of page 99 that there could exist some extremely large prime number above which the theorem breaks down for all larger primes.

The proof of the Mordell conjecture was announced in the summer of 1983 by a 29 year old West German mathematician, Gerd Faltings. But does it really open up a path towards a final proof of Fermat's last theorem? No-one knows for sure, just yet. In any case the proof of the Mordell conjecture was a major mathematical event, and Harvard mathematicians even interrupted their relaxed summer schedules to stage an intense series of seminars on its implications.

As regards other topics covered in the text we are pleased to announce the arrival of three more Mersenne numbers (see Chapter 2) the 28th, 29th and 30th members of that famous series of prime numbers of the form $2^p - 1$. Each was, at the moment of its discovery, the world's largest known prime number, and each generates a new perfect number via the formula $2^{p-1} \times (2^p - 1)$, see Chapter 9. These three new record breaking primes $2^p - 1$ are for $p = 86,243$ (discovered 1982), $p = 132,049$ (1983) and $p = 216,091$ (1985). The last is hot off the press, or rather the radio since I heard about it over my car radio on my way to work just last week. In this sense it is unconfirmed, since it may have become garbled in its passage from its discoverer to the Los Angeles Times (where the radio news reporter claimed to have found it) to the radio station to me. If true, however, it contains 65,050 digits and leads to a perfect number (only the thirtieth such number known) with no less than 130,100 digits. For those interested in the *last* digit of perfect numbers (see the sequence of sixes and eights on page 75) the three new 'perfects' add the contribution 868 to the pattern and thereby break up the run of seven sixes which arose from the 21st to the 27th perfect number inclusive.

The research on the Fermat 'primes' F_n of page 15 has continued apace but still no actual primes of this form larger than F_4 have surfaced. Many more F_n have been tested (with F_{9448} being the largest yet successfully tested) for primeness and there are now seventy-five F_n with n larger than 4 known to be composite (i.e. not prime) and none known to be prime. The number F_{20} remains the smallest Fermat 'prime' whose character (prime or composite) remains unknown. F_{14} is known to be composite even though none of its factors has been found. How can that be possible, you ask? Unfortunately that must remain a story for a future book, but methods are indeed available which can establish that a number must have factors even though none is known. Of all the F_n successfully tested, only F_5 to F_8 have been completely factored.

On a different front more progress can be reported on the seemingly never-ending computation of pi. The ten millionth decimal place was reached in 1983 by a Japanese research group using thirty hours of computer time. The number shows every indication of being normal (in the sense of Chapter 18). A second 'run' of seven consecutive threes (starting at the 3,204,765th decimal place) has been unearthed (see page 161) as well as runs of the same length of every digit except 2 and 4. In accord with the expectations for a normal number we are also able to extract such other hidden secrets as the first seven digits of e (see page 175), namely 2718281 (at the 1,526,800th decimal place) and the first eight digits of the square root of two, namely 14142135 (at the 52,638th decimal place). Plans are already underway to proceed to the one hundred millionth decimal place and the billionth decimal place may well be in sight before the decade is out. In accord with this mania for generating strings of hopefully patternless digits I note that factorial one million ($10^6!$, see page 26 for its definition) has also been calculated. It contains 5,565,709 digits and filled out a stack of standard computer printout paper over five inches high.

Finally I find that the number

$$10^{10^{10^{10^{34}}}}$$

on page 30 is no longer, by any means, the largest which has ever served any useful purpose in mathematics. In a recent proof, Dr. Ronald Graham of Bell Laboratories, New Jersey, established that a certain quatity involved could not possibly exceed a number which he called G_{64}. Now G_{64} is formed by *starting* with G_0 which is

$$3^{3^{3^{3^{3}}}}$$

and proceeding to calculate G_1 as 'G_0 to the power G_0', G_2 as 'G_1 to the power G_1' and so on all the way out to G_{64}. The number G_1 is already about '10 to the 10 to the 10 to the 12th' and G_{64} is a number so large that there is just no compact way of writing it down at all. Mind you, just how 'useful' its purpose in mathematics is can perhaps be queried since the number actually sought in this same problem is generally believed to be six (!).

APPENDIX 1; from Chapter 2, Page 11

Euclid's proof that the number of primes is infinite involves numbers of the form

$$(1 \times 2 \times 3 \times 5 \times 7 \times 11 \times \cdots \times N) + 1$$

where the part in brackets contains all the primes up to and including N (which is also a prime). Euclid's argument was that this number could not be exactly divisible by any prime up to and including N (since division by any of these always produces the remainder 1), so that it must either be a new and larger prime number itself, or be divisible by a prime larger than N. Here we discuss the question of how often the number set out above is itself prime. Surprisingly, perhaps, very few are once we get beyond the smallest examples $N = 2$, 3, 5, 7, and 11. In fact, after $N = 11$, only four more of these so-called *Euclidean* primes exist up to $N = 1031$; they are for $N = 31$, 379, 1019, and 1021. In view of the comparative rarity of these special types of prime, the occurrence of a pair of twin-primes just beyond 1000 in the series for N is a delightful surprise.

Recently, anthropologist Reo Fortune has suggested that if P is the smallest prime number which is strictly greater than any particular (prime or non-prime) Euclidean number of the above form, then the number defined by

$$P - (1 \times 2 \times 3 \times 5 \times 7 \times 11 \times \cdots \times N)$$

is *always* prime. To illustrate this, consider the number

$$1 \times 2 \times 3 \times 5 \times 7 \times 11 \times 13 \times 17 = 510{,}510.$$

The next prime larger than the corresponding Euclidean number 510,511 is 510,529. The 'fortunate number' for this case is therefore $510{,}529 - 510510 = 19$ and it is prime. The complete sequence of 'fortunate numbers' begins

$$2, 3, 5, 7, 13, 23, 17, 19, 23, 37, 61, \ldots$$

and to date all such numbers which have been tested are prime, although no general proof of Fortune's conjecture has been given.

We now note, however, that Euclid need not have used the special Euclidean form with a '+1' on the end; he could equally well have used a '−1' to prove his point. Thus, the numbers

$$(1 \times 2 \times 3 \times 5 \times 7 \times 11 \times \cdots \times N) - 1$$

are of equal interest in the present context. For values of N up to 307, only 6 prime numbers of this form are known. They are for $N = 3$, 5, 11, 13, 41, and 89.

Two other kinds of number could also have been used for establishing that primes go on forever. They are the forms $N! + 1$ and $N! - 1$ where $N!$ stands for

$$1 \times 2 \times 3 \times 4 \times 5 \times 6 \times \cdots \times N$$

in which *all* the integers less than or equal to N are included in the product. The numbers $N! + 1$ and $N! - 1$ are not exactly divisible by any prime less than or equal to N (since, once again, a remainder of 1 is left in each case) so that each must either be a new prime or be divisible by a prime larger than N. We now ask how often these numbers are themselves primes. It happens that prime numbers of the form $N! + 1$ and $N! - 1$ are rather more common than their Euclidean counterparts although there are still far more composites than primes in the series. For values of N up to 230 the number $N! + 1$ has been found to be prime when

$$N = 1, 2, 3, 11, 27, 37, 41, 73, 77, 116, \text{ and } 154.$$

For $N! - 1$ I am aware only of the findings for values of N up to 100. In this range, prime numbers result when

$$N = 3, 4, 6, 7, 12, 14, 30, 32, 33, 38, \text{ and } 94.$$

In all cases there are indications that the primes get rarer as N increases, and it remains an open question as to whether primes can be found for any or all of these number forms to arbitrarily large values of N.

APPENDIX 2; from Chapter 5, Page 33

In the prime number 'race' between primes of the form $4n + 1$ and $4n + 3$, that is the two series which begin

$(4n + 3)$:   3, 7, 11, 19, 23, 31, 43, 47, 59, 67, 71, 79, 83
$(4n + 1)$:   5, 13, 17, 29, 37, 41, 53, 61, 73, 89, 97

it was noticed as long ago as the year 1853 that the top row seems persistently to contain more primes than the bottom (when all the primes are counted up to a certain limit) and that this imbalance is maintained at least up to many thousands. In 1914, however, it was theoretically established that there must be an infinite number of integers below which the bottom row has more members than the top. This implies that, although the top row may be ahead for most of the time, there must be places where it loses the lead and, moreover, these places must occur from time to time all the way out to infinity. Because of this unusual situation there has been much interest in pursuing this particular race by computer to see when these 'cross-overs' really do take place. The study has now been accomplished all the way up to 2×10^{10} (which is twenty billion). It is found that in this range there are six separate regions where the bottom row 'becomes competitive' and actually takes the lead from time to time. Outside these regions the top row is always ahead.

We have to go all the way up to 26,861 before the bottom row goes ahead for the first time, and even then it is for a paltry single term. For a fleeting moment the bottom row contains just one more term than the top. Above 26,862, the top row pulls level again and forges back into the lead. This lead then continues all the way up to 616,841 when, for a period of nearly 17,000 integers (in fact up to 633,798) the bottom row gets back into the race, actually taking the lead at times and even getting as far ahead as 8 when we reach the number 623,681.

The other places where the bottom row gets ahead from time to time are near twelve and a half million, and near 1 billion, 6 billion, and 18 billion. The largest such region (by far) is the last. It begins at 18,465,126,293, and for the subsequent 568 *million* integers the bottom row is dominant, leading nearly all the time, and reaching a point near 18.7 billion where it has no less than 2,719 more terms than the top row. Inevitably, however, (or so it seems) the top row finally regains a persistent lead once more at 19,033,524,538.

Before this numerical work was completed in 1978, mathematicians had generally supposed that the percentage of numbers for which the bottom row was winning would get smaller and smaller as ever-larger numbers were reached. However, the block of numbers between 18 and 19 billion which favors the bottom row is, as a percentage of the number reached, by far the largest of any of the six regions yet encountered (up to 20 billion). This leaves us wondering whether it may be necessary for these mathematicians to have second thoughts. Could there really be a hope for a truly competitive underdog after all?

APPENDIX 3; from Chapter 11, Page 99

As an encouragement to those who think that problems which have evaded solution for generations must have answers so complicated as to be beyond the comprehension of the average non-mathematician, I offer the following gem. It dates from the middle of the nineteenth century and concerns a game of joining dots by straight lines on a piece of paper. If I draw an arbitrary pattern of dots like that in Figure 9a, and then join *every* dot in all possible ways to *all* the other dots, I find that some of the lines have only two dots on them while others have more. Is there any way (other than having all the dots in a straight line) that I can make a dot-pattern which will generate only lines with more than two dots on each.

The 5 dots in Figure 9a give rise to 6 straight lines, but 4 of them have only two dots on. We can improve the situation by adding a strategically placed sixth dot as shown in Figure 9b. This generates only 1 more line but fills in a third dot on two others, giving us 7 lines in total and now only 3 with two dots on. This is a small improvement but, no matter how hard we try, it seems impossible to get rid of all the two-dot lines. In fact, so difficult is it that well over a century ago the suggestion was made that it is impossible. But since the rules of the game allow you to play with any finite number of points (billions, trillions, or even more) how could anyone possibly prove such a thing?

Several generations of mathematicians tried, and failed, and there the situation remained until a moment of insight arrived (a moment of 'aha!' as that 'Dean of Recreational Mathematics' Martin Gardner would call it). The 'aha!' proceeds as follows. Consider an arbitrary dot in any pattern of dots. Some lines go through it, others don't. Of the latter, one line will come closer to the point than any other. Let this closest approach distance be called d. Now let us move to another point. The same argument gives us another d-value which may be

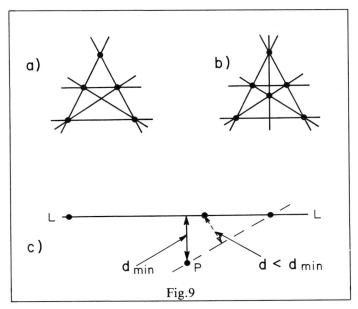

Fig.9

larger or smaller than the first. After testing every single point in the pattern there will be one d-value which is smallest of all - call it d_{min} (with min standing for minimum). Concentrate on this special point P for a moment: in fact, let us draw it, and show the line L which approaches it most closely. This is done in Figure 9c. How many points are on this 'closest line of all'?

"How should I know?" I hear you say, you haven't drawn much of the total picture. But the beauty of the 'aha!' is that I don't need to. Just suppose that this special 'closest line of all' had more than 2 dots on it. I will put three dots on it in the figure to emphasize the point. If it does, then we can always pick one of these three points (the middle one) and construct a d-value smaller than d_{min}, as shown in Figure 9c. But this has to be impossible since we have already stated that d_{min} is the smallest d-value which exists among all the dots of the pattern. It follows that the line L, with $d = d_{min}$, *can have only 2 dots on it*. The conjecture is therefore proven. To be completely truthful we do have to include the possibility of there being more than one point with the same minimum value of d, but this situation (which actually occurs in Figure 9a) can be covered by the simplest of extensions of the argument.

Do you not find this proof a thing of beauty? Unfortunately, it does not follow, by any stretch of the imagination, that such astoundingly simple proofs can be found for many of the other unsolved mathematical problems of this world - but perhaps it does give just a ray of hope that there might somewhere be an 'aha!', even for the likes of Fermat's last theorem.

APPENDIX 4; from Chapter 15, Page 135

We seek here a proof that the square root of any number which is not an exact square of an integer is irrational. Let us pick a particular example at random, say $\sqrt{22}$. The proof below is obviously reproducible for any other arbitrarily chosen example. We first assume that the answer *can* be written as a fraction, say n/m, which has already been reduced to its simplest form (by dividing top and bottom by any common factors). By a simple test we can now easily find neighboring integers such that the smaller is less than, and the larger greater than, the required answer n/m. For the case of $\sqrt{22}$ these integers are 4 and 5, whose squares (16 and 25) are respectively smaller and larger than 22. It follows that the number $5-(n/m)$ must be smaller than 1. We now define two other numbers n' and m' as follows:

$$n' = n(5 - n/m)$$

and

$$m' = m(5 - n/m).$$

From what we have said, these numbers must be respectively smaller than n and m. We also note that they have exactly the same ratio as n and m; that is

$$n'/m' = n/m$$

Using a little first-year algebra we can simplify the two defining equations for n' and m' to find

$$n' = 5n - (n^2/m^2)m = 5n - 22m$$

and

$$m' = 5m - n,$$

where we have used the fact that (by its definition) $n^2/m^2 = 22$. It follows that n' and m' must be integers (since m and n are). But they are also smaller than m and n, and have the same ratio as n/m. This is not possible since we started by saying that n/m was already the required fraction involving the smallest possible integers. The only conclusion which can be drawn is that our initial assumption must be wrong. That is, the square root of 22 just cannot be represented by any rational fraction, and the proof is now complete.

APPENDIX 5; from Chapter 21, Page 194

The proof that 'the number 2 raised to the power of *any* transfinite number is a larger transfinite number', is one of the simplest and most pleasing proofs in all of mathematics. First one establishes that the total number of ways of choosing zero, one, or more objects from a set of n is 2^n. As an example, if there are three objects in the set called A, B, C, then the $2^3 = 8$ different ways of choosing are

A, B, C, AB, BC, CA, ABC, and 0.

It follows that, for an infinite set with ∞_n members, combinations can be picked out in 2^{∞_n} different ways.

Suppose that we try to place all the ∞_n members of the set, starting with A, B, C, D, E, ... etc. in one-to-one correspondence with the various combinations that can be constructed from them. Note that we are no longer trying to count anything with the integers 1, 2, 3, ..., we are strictly searching for a one-to-one correspondence between the ∞_n members of the set and the 2^{∞_n} combinations which can be arranged by grouping the members together in all possible ways. Let us start the matching procedure as follows:

A	A,B,G,H,....
B	C,F.
C	C,D,F,J,....
D	A,B,F,J,....
E	C,D,G,H,....
F	A,C,F,G,....

Some of the combinations (like that opposite B above) will be finite; but nearly all will be of infinite extent. The matching arrangement is such that each matched pair is one or the other of two kinds; either the member in the left column is contained in the combination opposite it (as happens for A, C, and F) or it is not. To make this clearer we shall 'star' the members which are contained in their paired combinations, and bring the paired member to the front of its combination grouping. In this way we get:

$$A^*$$ $$A^*,B,G,H,....$$
B   C,F.
$$C^*$$ $$C^*,D,F,J,....$$
D   A,B,F,J,....
E   C,D,G,H,....
$$F^*$$ $$F^*,A,C,G,....$$

and so on.

Consider now the particular combination made up of all the *unstarred* members of the set. There must, of course, be such a grouping somewhere down the list on the right-hand-side. Should it be matched with a starred or an unstarred member in the left hand column. Suppose first that it is matched with a starred member. Then, by definition, the first member on the right hand side combination must also be starred. This cannot be the case since no member of the combination of concern is starred. Therefore we cannot pair this wholly unstarred combination with a starred member of the set. Alright, then let us try to match it with an unstarred left-hand-side member. It (again by definition) would require that this particular unstarred member was not contained in the matching combination on the right. But we have said that *all* the unstarred members are in this particular grouping of interest. Once more we reach a contradiction. The only conclusion is that this particular combination containing all the unstarred members cannot be successfully matched at all. In other words there is no one-to-one correspondence between the ∞_n members of the infinite set and the 2^{∞_n} combinations of members which can be formed from it. The latter is always the larger, and therefore 2^{∞_n} is always a larger infinity than ∞_n.

BIBLIOGRAPHY

A selection of books of a primarily non-technical nature for further reading.

ADLER I 1960 *The New Mathematics* (New York: Signet)

BEILER A H 1964 *Recreations in the Theory of Numbers* (New York: Dover)

BECKMANN P 1974 *A History of Pi* (New York: St Martin's Press)

DAVIS P J 1963 *The Lore of Large Numbers* (Westminster, Maryland: Random House)

GUILLEN M 1983 *Bridges to Infinity* (Boston: Houghton Mifflin)

LOWEKE G P 1982 *The Lore of Prime Numbers* (New York: Vantage)

NIVEN I M 1961 *Numbers Rational and Irrational* (Westminster, Maryland: Random House)

ORE O 1948 *Number Theory and its History* (London: McGraw-Hill)

—— 1967 *Invitation to Number Theory* (Westminster, Maryland: Random House)

RUCKER R 1982 *Infinity and the Mind* (Basel: Birkhauser)

SONDHEIMER E and ROGERSON A 1981 *Numbers and Infinity* (Cambridge: Cambridge University Press)

STEEN L A 1978 *Mathematics Today* (Berlin: Springer)

STEWART I 1975 *Concepts of Modern Mathematics* (Harmondsworth: Penguin)

For the potentially more serious student an excellent introduction to number theory can be found in

DAVENPORT H 1952 *The Higher Arithmetic* (New York: Hutchinson House)

while a sobering glimpse of the mountains which still remain to be climbed in the subject can be obtained from

GUY R K 1981 *Unsolved Problems in Number Theory* (Berlin: Springer)

INDEX